Confessions of a SCILLY BIRDMAN

Confessions of a SCILLY BIRDMAN

David Hunt

Foreword and postscript by Bill Oddie

CROOM HELM
London · Sydney · Dover, New Hampshire

© 1985 David Hunt
Line drawings by Bryan Bland
Croom Helm Ltd, Provident House, Burrell Row,
Beckenham, Kent BR3 1AT
Croom Helm Australia Pty Ltd
Suite 4, 6th Floor, 64-76 Kippax Street,
Surry Hills, NSW 2010, Australia

British Library Cataloguing in Publication Data

Hunt, David
    Confessions of a Scilly birdman.
    1. Hunt, David   2. Ornithologists–Great
    Britain–Biography
    598'.092'4   QL31.H/
    ISBN 0-7099-3724-5

Croom Helm, 51 Washington Street, Dover,
New Hampshire 03820, USA

Library of Congress Cataloging in Publication Data

Hunt, David, 1934-1985.
    Confessions of a Scilly Birdman.

    I. Hunt, David, 1934-85.   2. Bird watching–England
    –Scilly, Isles of.   3. Birds–England–Scilly, Isles
    of.   4. Ornithologists–England–Scilly, Isles of–
    Biography.   I. Title.
    QL31.H86A33   1985   598'.07'234   [B]   85-7847
    ISBN 0-7099-3724-5

Typeset by Leaper & Gard Ltd, Bristol, England
Printed and bound in Great Britain

Contents

List of Illustrations	7
Foreword *Bill Oddie*	9
Introduction	11
Chapter 1 Recollections of a Fledgeling	13
Chapter 2 The Juvenile Spreads his Wings	27
Chapter 3 Years of Immaturity	43
Chapter 4 The Breeding Period	61
Chapter 5 Post-breeding Dispersal	75
Chapter 6 The Adult Birdman Emerges	113
Chapter 7 Settling into a New Habitat	131
Chapter 8 Leader of the Flock	145
Chapter 9 Adventures in the Bird Trade	161
Chapter 10 Present and Future Status	171
Postscript *Bill Oddie*	174

LIST OF ILLUSTRATIONS

Line drawings by Bryan Bland

Foreword photograph: Bill Oddie
 1 Fledgeling with parents
 2 Fledgeling and friend
 3 'But that's not a Tulip-tree, Grandpa, it's a Magnolia'
 4 'I became aware of its hooked beak and glaring yellow eye'
 5 'The elusive Silver-washed Fritillary'
 6 Spoonbills from the train
 7 A Great Spotted Woodpecker on the school lawn
 8 Marsh Harriers at the nest
 9 Stone Curlews on the Norfolk Brecks
10 Junior Cricket
11 The elusive Solitary Sandpiper
12 Familiar waders on Cley Marsh
13 My first Hoopoe
14 'Baggers'
15 An early rarity – Pectoral Sandpiper
16 A heavy crop of Waxwings
17 Richard Richardson
18 Migrant watching in Norway
19 Temminck's Stints
20 'I also came to sample the edible qualities of Cormorants and gulls'
21 North Sea stowaway
22 Me and my army mates – that's me on the right
23 Birdwatching army style
24 A youthful artistic endeavour
25 Refreshments in Andorra
26 Bee-eaters overhead
27 The art school image
28 A hot night in the Boheme Club
29 My two early loves
30 Kites over the Rhine
31 White Storks at the nest
32 Après Stork
33 The stork I never saw
34 Marianne became a Bunny Girl
35 Nicholas at the Round Pond

36　Roof-top Shelducks
37　Top billing in the West End
38　'A Snowy Owl materialised'
39　A Gyr Falcon for Christmas
40　Unprecedented arrival of Hoopoes
41　Early morning at Men-a-Vaur
42　St Agnes' famous lighthouse
43　'It's a yankee cuckoo!'
44　White's Thrush – the fantasy realised
45　'That bird belongs on Bryher'
46　Singing Lesser Whitethroat
47　Bluethroat from the bathtub
48　Britain's first Parula Warbler
49　St Agnes Observatory
50　Settled in at the Blockhouse
51　The credible Long-billed Dowitcher
52　'Ah Hunt – conserving your energy I see'
53　A Wryneck at last
54　As photographed by Martin King
55　Northern Waterthrush – definitely not a BBRC plot
56　'Sorry – my mistake!'
57　Spot the real Buccaneers
58　Sunday 'Seabird Special'
59　Cyril Ellis
60　Semi-western Sandpiper?
61　First (and last) exhibition at Plymouth Athenaeum
62　'Red Sails' and the Sunset Syncopators
63　Shearwater Special
64　Some familiar faces?
65　St Mary's first-ever traffic jam
66　A lucky shot of a Nighthawk
67　The momentous boat-trip
68　Robert Allen
69　The Greater Lesser Yellowlegs
70　Products of an easterly wind
71　Some of the Grey-cheeked Thrushes
72　Semi-palmated Plover (left) attending Hilda Quick's funeral
73　Rock Thrush – reality from an illusion
74　A familiar Stonechat
75　Tresco Myrtle Warbler
76　Snow Bunting or No Bunting?
77　That's all folks!

Foreword

I'm assuming that if you are reading this, you are a bird-watcher. There can't be many bird-watchers who haven't heard of the Isles of Scilly; and there's probably not a lot who haven't actually been there at one time or another. At the very least, you must have browsed through the 'small ads' at the back of *Birds* magazine or *British Birds*. In which case, you'll certainly know the *name* of David Hunt. Maybe you've just seen his little advert in its nice square box: 'Enjoy the Holiday of a Lifetime, with the Man on the Spot' (David Hunt). Or maybe you've even spotted the man himself. Maybe you've even spoken to him, though perhaps you didn't realise who he was at the time. *I* didn't know who he was when I first met him, and if I hadn't seen him on and off for the last fifteen years, I don't think I'd recognise him if I met him again today. His appearance has changed quite a bit. Back in the late 60s (or was it the early 70s?) he resembled a rather morose overweight Beatle (a 'She loves you yeah yeah' Beatle – not an insect). Nowadays he's slimmer, better groomed, altogether more dashing, I think more confident and – dare I say it? – happier. In fact he even seems younger. How he's come to arrive at this enviable condition is told in this book.

I have known David for well over a decade, and over more recent years I've also had the great pleasure of getting to know many more of the 'establishment figures' of British Ornithology. *Names* I only used to see heading abstruse papers in *BB*, recording rarities or publishing books have now become real live *people*. To me this is *the* great delight of being involved in the 'bird-world' – the fact is, bird-watchers are wonderful company; and they make very good friends. David Hunt is both, and not least because his experience and conversation are by no means limited to birds – in fact there are many days he refuses to talk about birds at all. He is also an all-round naturalist, a horticulturalist, an artist and a jazz musician. At various times he has saved my ailing pot plants; insulted my paintings; sat in an audience and suffered my saxo-

Bill Oddie

phone playing; and got his own back by drowning out my drumming with his trombone (no mean achievement).

So, this book is not only about birds and bird-watchers (though there's plenty of both). David calls it his 'confessions', and the title is justified. You will perhaps be shocked, but surely not offended, as he shamelessly admits to stealing trees, eating birds and a sordid past of sex, drink and rock and roll. He also tells of island feuds and scandals that give 'Dallas' a run for its money and – most alarming of all – he even confesses to having once 'strung' a Gull-billed Tern!

David Hunt is in fact not *a* Scilly Birdman – he is *the* Scilly Birdman. You know the name, you may have seen the face, you may have heard the voice – NOW READ THE BOOK ... and discover the person. Even if you've never heard of him – you'll enjoy getting to know him.

Bill Oddie

Introduction

How does one come to be known as a 'Birdman' as opposed to being just another bird-watcher, or, as many style themselves today, birder? Such subtle distinctions are not easily grasped by the uninitiated, but the title of Birdman is usually linked with some location as in 'Birdman of Alcatraz', or, in my case, 'Scilly Birdman' – a sobriquet that some might say I deserve in more ways that one!

In Shetland the name of Bobby Tulloch comes immediately to mind, and at Cley in Norfolk one thinks of Billy Bishop (although there have always been other contenders for that crown, foremost among them one of my own mentors, the late R.A. Richardson).

The Birdman of Alcatraz, of course, was a unique and extraordinary person – a convicted criminal who, during his incarceration in the prison on the infamous island in San Francisco Bay, became a bird expert in his own right, and eventually the subject of a film starring none other than Burt Lancaster. His story has no bearing whatever on mine, unless it be argued that I too am a prisoner on a lonely island. But in my case the captivity is a voluntary one and if I cannot escape the Isles of Scilly it is because of my love for the islands, and my good fortune in having found a livelihood among them.

Just when I assumed the title of Scilly Birdman it would be hard to say, but throughout the twenty years I have lived in the islands, birds and bird-watching have been an important part of my life, and during that time I have been lucky enough to be able to earn a modest living through talking about them, photographing them and showing them to summer visitors. Many well-heeled holidaymakers have envied me my open-air existence, free as it seems to be from all the daily stresses of mainland living; but what they probably do not recognise is that, having proved to be totally unsuited to conventional kinds of work, and certainly incapable of holding down a steady job for any length of time, I am only here through default and have only become what I am as the result of an extraordinary chain of events.

The story that follows relates how, after a series of false starts, embarrassing failures and a few happy coincidences, I eventually became the Scilly Birdman. It starts with my days as a fledgeling in Devon, followed by a juvenile stage in Norfolk, immaturity in London with migrations to and from Europe, nesting in Germany with post-breeding dispersal at home and abroad, and finally adulthood (though never probably maturity) as a fully-fledged birdman in the Isles of Scilly.

1. Recollections of a Fledgeling

I remember nothing of my time in the egg, which is not all that surprising, and very little of my time in the nest. In fact, I was born by Caesarean section on 8 May 1934 in the Royal Naval Hospital at Devonport. My mother, who was forty-three at the time (I was to be her only child), remembers the events a lot better than I do, which is also not very surprising!

1. Fledgeling with parents

I suppose a lot of our childhood memories are partly what we have been told by our parents, and I certainly have to rely on them for my earliest recollections, though of course some of them stem from my own mind as well. What follows now is an amalgam of memories: some are my own, others my parents'.

We lived in the village of Newton Ferrers in South Devon, situated not far from Plymouth as the crow flies, being near the mouth of the River Yealm, which is the next estuary to the east. My father O.D. Hunt (his first name Owen was rarely used except by family and closest friends) was a marine biologist by profession, and all-round naturalist by inclination. Formerly an assistant naturalist at the laboratory of the Marine Biological Association in Plymouth, father at the time of my arrival had recently become head of a research laboratory doing experimental work for a firm producing marine paints, with a brand new establishment purpose-built on the shore of the River Yealm. Our house was perched on the hillside overlooking the river, among giant pine trees in a garden full of interesting plants and shrubs, for gardening was among my father's special interests. To reach the house it was necessary to climb several flights of steps, but it was worth it for the superb view of the river it afforded. Not that any of this impressed me at the time – for me it was just the place where I lived, and I took it all for granted.

My earliest personal recollections of Newton Ferrers are of being taken for walks, in company with children of my own age from another family, along country lanes and in the woods and fields surrounding the village. Even in those days I was very conscious of the birds, wildflowers and butterflies and no doubt I was greatly encouraged in this by Father, as I always called him, right from the start. Our house contained plenty of reading matter, especially works of a scientific nature, and I loved to pore over the illustrations in natural history books even before I learned to read

2. Fledgeling and friend

them for myself. Books which I especially recall were *Before the Dawn of History* and *Animated Nature*, both full of graphic illustrations of exotic creatures, especially dinosaurs, which I found especially fascinating. No wonder, then, that when asked by relations or friends of the family that perennial question 'What do you want to be when you grow up?' my stock reply in those days was 'A palaeontologist'! I doubt if I could have spelt it (I'm still not sure), but I knew what it meant and I said it quite sincerely I'm certain. From this you will deduce that I was a precocious little beast and another instance of this is recalled by my mother, who inconsistently I always called Mummy! We were paying a family visit to London where my grandparents still lived and Grandpa, who no doubt wished to contribute to the store of knowledge already building up inside my head, showed me what he called a 'Tulip-tree' in a neighbour's garden. 'But that's NOT a Tulip-tree, Grandpa, it's a MAGNOLIA,' replied the scandalised toddler. Whether I was right or wrong is not recorded but what emerges from this anecdote is that I was

3. 'But that's not a Tulip-tree, Grandpa, it's a Magnolia'

not afraid to correct my elders and betters even at that early age. This annoying trait is one that I have passed on to my own children, as I have since discovered through painful experience!

I am not sure when my special interest in birds started but Father told me that I was able to distinguish Bramblings from Chaffinches coming to our garden bird-table at a date many years ahead of my personal first recollection of seeing a Brambling! It was a close-range view of a Buzzard which first really 'turned me on' to birds as wild creatures, and this happened while I was a pupil at St Faith's – a preparatory school evacuated from Cambridge to Ashburton, a sleepy little town on the southern fringe of Dartmoor. I was sent there at the tender age of seven as a boarding pupil during the early years of the Second World War. There I was relatively safe from the attentions of Hitler's bombers which were almost nightly making air-raids on Plymouth – and Newton Ferrers, being directly in their flight path, received stray bombs from time to time.

To return to the Buzzard – I had disturbed it from its prey, a bloody young rabbit which had been abandoned in the ditch, but in the brief moment before the Buzzard flew I became aware of its hooked bill and glaring yellow eye, seemingly daring me to come

4. *'I became aware of its hooked beak and glaring yellow eye'*

any closer. These features symbolised a wildness that I had not previously recognised in everyday garden birds – though Buzzards were, and still are, a regular sight in South Devon, I had never seen one at close range before and that fierce image remained in my mind's eye long after the Buzzard had flapped away and out of sight.

No doubt to make life at boarding school more bearable my parents visited me as often as they could at weekends and, on one of those occasions, Father let me have an old pair of field-glasses (the term binoculars was little used then). He also gave me his two volumes by T.A. Coward, *The Birds of the British Isles and their Eggs*, which until then had been the standard work for bird identification. Father had just purchased the new *Handbook of British Birds*, published by Witherby, so parting with Coward was not the sacrifice it might have been, and it certainly helped to make days at St Faith's a little less miserable. We used to be taken for walks in the country almost daily, but invariably in pairs and in 'crocodile' formation which was rarely relaxed. However, on one memorable occasion I discovered a nest with eggs, wedged among stems of meadowsweet, which by means of Coward's book I identified as that of a Whitethroat – my first unaided ornithological discovery!

School holidays came mercifully from time to time and, though I longed for them probably no more fervently than any other pupil at St Faith's, they offered not only escape from the awful restraints of boarding-school life, but opened up unlimited opportunities for adventure and discovery. Together with a handful of friends, and older boys from the village, I roamed the woods and fields in search not only of birds' nests (strangely we rarely took eggs) but also of fragments of shrapnel and other souvenirs of the air-raids which were of almost nightly occurrence. Most people had a bomb-shelter of some sort to retire to when the sirens went, but our house had a specially reinforced ceiling in the kitchen, and for added protection I slept in a camp-bed under the kitchen table. Whether or not it would have survived a direct hit we luckily never had the chance to find out, but I don't recall ever feeling afraid – only being excited by the sounds of anti-aircraft fire and the occasional scream of a falling bomb. Father was not required to join up for active service (he had already done his share as a very young stretcher-bearer during the previous war) but acted as local Air-raid Warden, so perhaps this gave me an added sense of security. It also sometimes helped to provide an extra special souvenir, like the unexploded incendiary bomb, which for years was one of my most treasured possessions!

During the hours of daylight, bombing raids were only the remotest possibility and we boys had the freedom of the countryside. I soon got to know every corner of the woods and fields around the village and became familiar with all the common birds, learning to recognise their songs and calls as well as their appearance. Other wild creatures like rabbits and red squirrels were also in abundance though the more nocturnal species, like foxes and badgers, went unobserved.

It was about this time, during the early forties, that I also became fascinated by butterflies and moths. One of my older friends, Trevor, who must have been at least twelve or thirteen and therefore almost twice my age, though I had more in common with him than any of the others, had already started a collection of butterflies. This probably comprised no more than about thirty specimens, representing perhaps a dozen species, all pinned out neatly in a cork-lined box. They were mostly of the Peacock and Small Tortoiseshell variety – but among them was a superb Silver-washed Fritillary which somehow gave the collection real class! I was very envious of Trevor's achievement and it was one which I desperately wanted to emulate, though of course I knew I could not surpass it. However, a net was soon manfactured from a hoop of galvanised wire and some white muslin attached to a bamboo cane, and a killing-bottle to dispatch the unfortunate insects was simply made with an air tight glass jar and some laurel leaves. All I needed then was a setting board, and Father helped me make one of those. The specimen case I started off with came from a retired gentleman who had once collected butterflies during his days as a planter in India or Africa (it meant little to me then). His specimens were instantly rejected and consigned to the dustbin – somehow they seemed very dull and uninteresting compared to the splendours of a Silver-washed Fritillary!

5. *'The elusive Silver-washed Fritillary'*

Recollections of a Fledgeling

The neighbouring garden to ours had a couple of large bushes of Buddleia (nowadays often known as the 'butterfly bush') and as I was always allowed to go in there it was not long before I also became the proud owner of a collection of Peacocks, Red Admirals and the like. The Silver-washed Fritillary, though, proved a different matter for although one did appear from time to time about the Buddleia it always settled tantalisingly just out of reach, and at the first wild sweep of my net would glide away to safety in the overgrown and thus inaccessible woodland on the far side of the hedge. So for the time being I had to be content with building up my collection with Meadow Browns, Large Whites and assorted commoners. By the end of that summer my collection almost matched Trevor's though he had added a Purple Hairstreak in the meantime, and I still had no Silver-washed Fritillary!

The acquisitive instinct is inherent in most of us in some degree, I believe, and though it rarely becomes overtly competitive in adults, small boys have few inhibitions of this sort, and indeed such an establishment as St Faith's actively fostered the competitive instincts of its pupils in both classroom and on the sports-field. So it is not surprising that this sort of rivalry persisted in other areas of activity, notably in collecting manias. The principal manias of my school days were for military badges, which initially could only be acquired by purchasing them (an expensive business for parents) and bus tickets, for which one had to fight and grovel in roadside gutters during those character-forming walks along the Devon lanes. I had managed to assemble representative collections of both, but they paled to insignificance beside my butterflies. I knew that in this I would be unique among my fellow pupils, that none of them would be able to match, or could even attempt to match my collection of butterflies. In this area, if in nothing else, I would outshine them all when I returned to St Faith's.

The one who encouraged me most was my father, despite the fact that my mother seemed to disapprove. I suppose he saw it as an acceptable alternative to collecting birds' eggs, which I might otherwise have been tempted to do, and certainly there would have been plenty of opportunity for this if I had so desired. I now find pinning out butterflies in glass cases just as repugnant as stealing eggs but, in those days, nobody considered that taking a few insects was either cruel, or likely to deplete the populations of even the rarer species. As the son of a biologist it would have been strange indeed if I had not taken up a hobby like this, and Father of course could not have been better qualified to help me in the pursuance of it. Ironically, some twenty years later, my father was to become one of the leading figures at the start of the campaign to save the Large Blue butterfly from extinction in Britain – a situation undoubtedly caused partly by over-enthusiastic collecting, though not I am pleased to say by myself.

During the winter that followed I spent much of my spare time poring over a catalogue from Watkins and Doncaster, a firm which specialised in supplying the impedimenta necessary for collecting all kinds of natural history specimens from insects to elephants I dare say, but I was only interested in the pages concerning entomological pins, setting boards, nets and killing-bottles. My main Christmas present I remember was a proper butterfly-net, which folded up for carrying in the pocket, but which could be whipped out for instant deadly use if required. I also spent a lot of time musing over the pages of *British Butterflies* by F.W. Frohawk, learning all about the life histories of the insects which I hoped one day to collect. How eagerly I awaited the next summer and the opportunities it would bring to add to my collection!

In fact I did not have to wait as long as that because in the Easter holiday I discovered a habitat new to me, locally known as Spindle Hill. This was a scrubby area with grassy clearings, perhaps formerly pasture land, but rapidly being overgrown by invading brambles and saplings which spread out from the surrounding woodland. That spring was an exceptionally warm and sunny one and I spent many enchanted hours roaming the woods, especially the Spindle Hill area, where I found Dingy and Grizzled Skippers, Brimstones and Orange Tips to add to my collection – but the real prize of that holiday was my first Pearl-bordered Fritillary! Not as impressive as a Silver-washed maybe, but certainly a step in the right direction. I also stumbled one day upon a Pheasant's nest with eggs, but don't remember the slightest desire to take one – clearly my collecting instincts were channelled into a very narrow vein.

Although I must have spent nearly three years at St Faith's, the only things I clearly remember from those days are the moments of release from the dreary restrictions of institutionalised life, but I do recall well, and can visualise, probably because I have passed it many times since, that the school had taken over for its wartime premises the Golden Lion Hotel, complete with extensive garden and swimming-pool. Here we spent most of our young lives during term-time, either being crammed with the basic knowledge which one day might serve to provide us with a scholarship to one of the major public schools, or indulging in healthy pursuits like running round and round the lawn before breakfast in winter, to compulsory swimming lessons in summer. The only time we escaped these activities was if there was an epidemic of whooping cough or German measles and one was lucky enough to be confined to bed! There was also the blessed relief of half-term weekends, when parents were able to descend and remove their offspring for as long as a day and a half if they lived near enough. Luckily mine did, but we did not go back to Newton Ferrers – we would book rooms at the nearby Holne Chase Hotel and spend the few precious hours available exploring the countryside along the upper reaches of the River Dart.

That summer I collected my first Green Hairstreak, Dark Green and High Brown Fritillaries and, I also remember well, learned to recognise the shivering song of the Wood Warblers which frequented the enormous beech trees in the extensive woods of Holne Chase. Dippers and Grey Wagtails were a common sight along the river and one memorable evening we also had the good fortune to spot an otter with its young, as they gambolled in the shallows. My mother did not share our passion for wildlife and wild places but would come along just the same, preferring (while we were exploring) to spend her time reading or knitting in the hotel garden.

Father also occasionally managed to visit me for just a Sunday or Saturday afternoon and on those occasions we made shorter excursions into the lanes nearer the town, where otherwise I was always restricted to the dreaded crocodile, and on one such visit I was astonished to spot the unmistakable underside pattern of a White-letter Hairstreak as the tiny butterfly feasted on a bramble-blossom in the hedgerow. For some reason I did not have my net with me and, since there was no time to go back to school and fetch it, I had to resign myself to missing a chance of adding a specimen to my collection which would have been unique, since it was at that time previously unrecorded in Devon. My disappointment did not diminish during the days that followed and later that week I took an unprecedented risk, sneaking out of the school grounds during a swimming gala in which I was not taking part. Never an enthusiastic swimmer even in the warmest weather, I was not likely to be missed during the excitement of events round the pool – it was just a matter of escaping notice as I left and re-entered the school grounds. I do not remember anything of the circumstances that followed – only that at the end of the day I had secured the longed-for specimen, and that no one had had the slightest idea that I had absented myself from the school during that never-to-be-forgotten afternoon when I netted Devon's first White-letter Hairstreak! This event also led to my first appearance in print – a brief paragraph in the letters section of the *Western Morning News* recording the circumstances (though not the incriminating ones) of what I fondly imagined to be a milestone in scientific discovery. I also saw myself as a force to be reckoned with in literature as well as entomology – how could I have known better at the tender age of nine?

I don't suppose that St Faith's was much different from the many other similar establishments dotted about the country at that time, and who knows if at that age I would have adapted better to life in another prep school? But perhaps because I had red hair, freckles, and cried easily, or was it because I had unusual interests and was spoilt and conceited – whatever the reasons I found myself being increasingly picked on by the other boys. This was brought to a head when I was accused of stealing fruit from the kitchen garden (wrongfully of course) and became a scapegoat for the whole school to wreak vengeance upon, with the headmaster's approval! Since I was also failing to live up to expectations in the classroom and most of my teachers had it in for me as well, it is perhaps not surprising that I eventually implored my parents to take me away from St Faith's and find a new school where I could start with a clean slate, though I don't know what I expected.

What I got turned out to be a great improvement in every way, especially as Ravenswood, my new school, was in another remote region of Devon, this time on the southern fringe of Exmoor. I managed to integrate much better into the school life here and soon found that a moderate talent at games worked wonders with my fellow-pupils who did not regard my special interests as quite so unacceptable in one who could also play cricket and football. Another big advantage was that the head boy – a giant, but quietly spoken chap was also interested in natural history, as was his younger brother, so I had a friend in high places from the start, or if not a friend an ally at least, especially as the younger brother was the same age as myself. Academically I still could not live up to what was expected of me, but at least I held my own in most subjects even if I did not excel in any. Clearly the move to Ravenswood had been a fortuitous one, and this must have been reflected in my disposition because I soon became a cheerful extrovert, quite a different character from the quailing scapegoat I had been during my latter days at St Faith's.

It was not long before I was adding new and valued specimens to my butterfly collection. The still unpolluted countryside around Stoodleigh, the remote village not far from Tiverton where Ravenswood was situated, had habitats for such comparative rarities as Marsh and Heath Fritillaries, and at the start of one autumn term I found a colony of Brown Hairstreaks, later collecting their eggs from which I eventually bred out some fine specimens the following year. It was about this time that I started rearing butterflies and moths quite seriously with Father again proving an invaluable help since he was able to provide me with suitable cages and, more importantly still, look after the caterpillars or chrysalids at home during term-time when it would have been rather impractical to keep them all at school.

Journeys to and from Ravenswood were accomplished by train, the first leg of the trip being on the main line from Plymouth to Exeter, and it was along this route that I also kept in touch with my bird-watching for the railway line runs very close to two of Devon's major estuaries, the Teign and the Exe; it was from trains taking me to or from school that one day I saw a couple of Spoonbills, and on

6. *Spoonbills from the train*

more than one occasion a few Avocets. We also paid regular visits to Slapton Ley, Dawlish Warren and Wembury Bay during the winter holidays when butterfly hunting was out of the question. I became familiar with all the regular ducks, waders and other shore birds during these visits, and also learnt their characteristic calls and how to imitate some of them.

I was thrown very much upon my own company during school holidays from Ravenswood as Trevor was now finished with school and having to earn a living, and my other friends had all developed new interests, like girls, airguns and bikes, none of which appealed to me as alternatives to my own obsessions, which I was therefore obliged to pursue alone. Not that I missed their company particularly, having realised that, apart from Father, I knew nobody who shared my interests to the same degree. So I was content with just the companionship of my cross-breed terrier 'Pounce'. Together we would take off for long walks beside the river, or along the cliff path, and all the time my knowledge of the area and its wildlife gradually increased.

So, while most boys of my age were applying their energy to more conventional pursuits, I preferred to remain a loner. What made me feel even more apart from the other boys was discovering that some of them were using their airguns to shoot at garden birds – something which I just could not understand or even bear to contemplate – though equally to split on them would have been just as unthinkable! Instead, on the only occasion when I ever laid my hands on another boy's airgun, I successfully plugged him in the leg, and alienated myself even further from my former friends. No wonder that I have been an anti-shooter ever since.

Strangely, after those years of longing, I have no recollection of the day I finally caught my Silver-washed Fritillary – perhaps because before that happened I had become aware of migrant butterflies and moths. During the summer of 1944 Clouded Yellows, Painted Ladies, Convolvulus and Humming-bird Hawk Moths all found their way into my collection, and in 1945 I even added a supreme rarity – a Bath White, although in that year their rarity value was considerably diminished, for they turned up all over South Devon. Occasionally migrant moths could be found, too. I used to search for them around street-lamps at night, or by 'sugaring' – a process involving the spreading of a sticky mixture of molasses, beer and other secret ingredients on tree-trunks and fence posts at dusk, then collecting the intoxicated insects an hour or two later.

Two major events took place during my years at Ravenswood, the first of them being the end of the war. I remember this particularly, because 8 May 1945 also happened to be my eleventh birthday, and although the war had had nothing like the effect on me that it had on some boys at school, some having lost parents, others their homes, there was an immense feeling of relief and much celebration, even among those of us who had been least affected. For me the end of the war brought no radical changes, but it did mean a relaxation of travel restrictions and eventually the end of petrol rationing, which meant that Father and I could roam further afield in the pursuit of our interests, having until then been confined to our home county of Devon.

Not that my father was a Devon man, although he had chosen to settle in the county. East Anglia had been his childhood home and stamping ground as a young man, where he had clearly been profoundly influenced by the naturalist Arthur Patterson, whose books have since become regional classics. Father's early notebooks contain references to days spent in the company of Patterson himself, in pursuit of wildfowl and waders on the legendary Breydon Water as well as Benacre Broad and Lowestoft Denes. He was also associated with the eminent Suffolk ornithologist Claude Ticehurst, for whom he helped to prepare skins as a young man.

It is not surprising, therefore, that Father decided he would like me to go on for further education to Gresham's School at Holt in North Norfolk. Perhaps he hoped, through me, to relive those early days in East Anglia – certainly it gave him ample reason to return, after many years' absence, to what I suppose he regarded as his spiritual home. Anyway in the summer of 1946, we made the first of several holiday visits to Norfolk, ostensibly to look over the school at which it was hoped I would become a pupil, subject to my being able to attain the appropriate academic standard. This summer holiday was the second major event I referred to earlier, as it broadened my horizons which until that time had been narrowly confined to the county of Devon.

Before we leave Ravenswood altogether though, I should make mention of the severe winter of my final year there. The snow, which until then I had hardly experienced before, fell so heavily that our school term had to be abandoned due to shortage of fuel and other domestic supplies. This was not before I had managed to identify several Bramblings and Siskins among the birds which appeared in the woods around Stoodleigh, to say nothing of the Redwings and Fieldfares which were dying of cold and starvation all over the country. On a visit to the Exe estuary that winter Father and I also found large numbers of ducks, waders and other water birds which had been scattered dead along the shoreline, all victims of that exceptionally severe winter. This was the first time I became aware that wildlife too could experience mortality at the hand of nature, and this realisation made a lasting impression on me.

2. The Juvenile Spreads his Wings

Although it is almost forty years since we embarked on that momentous summer holiday in August 1946, there are still moments and scenes that I can picture in my mind to the present day, though I have had to piece together the actual circumstances by referring to old notebooks which fortunately have survived. In recent years I have been very lazy about making regular notes, except when on trips abroad where I always keep a daily journal of events and lists of species seen. Those early notebooks have proved invaluable to me in narrating this account of my boyhood enthusiasms, and demonstrate the great value of keeping records, however unimportant it may seem at the time.

7. *A Great Spotted Woodpecker on the school lawn*

The Juvenile Spreads his Wings

We travelled by car to Norfolk – quite an adventurous journey in those days. Father had relatives in Somerset, and I remember seeing for the first time the flat expanse of Sedgemoor near Langport, where my Uncle Jack managed a small farm. We searched unsuccessfully for Marsh Warblers while staying in this area for a couple of days before the next stage of our journey which took us all the way to Cambridge, where we stayed with an old colleague of my father's, a lecturer in zoology at the university. From here we went fossil-hunting on the chalk downs, and I was able to add both Chalk-hill and Adonis Blues to my butterfly collection. Cambridge sewage farm proved another happy hunting ground where Little

8. *Marsh Harriers at the nest*

Stints and a Ruff were among the waders – both new for me. The smell of sewage under treatment still evokes in me happy memories of that summer!

Although ostensibly this trip had a visit to Gresham's as its main goal, we only spent half a day there. My notebook makes no reference to my impressions of the school or our meeting with my future headmaster. 'Great Spotted Woodpecker on school lawn' is the only entry for that morning. Still, we must have realised that this would be the perfect place for me to continue my education for we found that the coastal marshes of Cley and Salthouse were only a few miles from Holt, the small market town where the school was situated. The surrounding countryside was in some respects not unlike the farmland I was used to in Devon, with woods and heathland as well. Whatever the school turned out to

9. Stone Curlews on the Norfolk Brecks

be like, I knew I would find plenty of distractions in my spare time.

Our next stop was the Norfolk Broads, and here Father was at last able to relive some of his boyhood memories. At Horsey Mere we were shown the nests of both Marsh and Montagu's Harriers, which had both only recently re-established themselves as breeding birds in Britain, thanks to the efforts of the landowners; elsewhere on the broads we were to see Bitterns, Bearded Tits and Garganey. I also found some caterpillars of the Swallowtail Butterfly which I fervently hoped to be able to rear in captivity. On leaving the Broads we headed for the Norfolk Brecks, an inland area of flinty heathland interspersed with pine plantations, where at dusk that evening we heard for the first time the eerie calling of Stone Curlews. Next day my diary recounts how we 'obtained excellent views of 12 Stone Curlews both on ground and in flight'.

It is not surprising that after that momentous holiday my main allegiance transferred from butterflies to bird-watching. Visiting such a range of new habitats had opened my eyes to the possibilities that travel offered and for the next few years our summer holidays were taken exploring other parts of southern England, usually ending up in Norfolk for the beginning of the autumn term.

Having been accepted as a pupil at Gresham's, though I did not succeed in winning a scholarship, the next year does not feature much in my memory, or my notebook – apart from the winter phenomena already described. My Swallowtails duly hatched out in the summer and I'm ashamed to say they were all destined for the killing-bottle and eventually the butterfly cabinet; although my interest in collecting was on the wane, it was by no means extinguished and a visit to Dorset that summer added Lulworth Skipper to my ever more comprehensive collection.

It was the autumn of that year, however, which I remember best as it was our first holiday at Cley, preparatory to my first term at Gresham's – the prospect of which I must admit filled me with tre-

The Juvenile Spreads his Wings

10. Junior Cricket

pidation! To be once more a new boy, among hundreds of other boys, most of them older and none of whom I knew previously, was not something I anticipated with a lot of enthusiasm. But meanwhile there was a week's bird-watching at Cley to look forward to, and I had no fears about that!

It was at Cley in September 1947 that I first discovered that other adults, beside Father and a few of his friends, actually took bird-watching as seriously as we did – in some cases even more seriously! We stayed at the George Hotel; I guess in those days there was no alternative, and as far as I remember everyone staying there was either a bird-watcher or a shooter! In fact there were only two shooters, Arnold and Higgins, but their presence in the dining-room each evening was sufficient to cause conversation about birds to be held in whispers or not at all. The other two guests I remember were Daukes and Harber, who did (when there

11. The elusive Solitary Sandpiper

was an opportunity to talk without being overheard) let us know a little about what birds were to be seen on the marsh. The big excitement that year was a Solitary Sandpiper which had been seen somewhere by someone, but whoever it was was keeping it a close secret for fear of Arnold and Higgins getting to know its whereabouts, which could only lead to one thing! Needless to say we never got to see the Solitary Sandpiper and whether it eventually found its way into Arnold's collection or not I don't know; but we did see our first Green and Wood Sandpipers, Spotted Redshanks, Curlew Sandpipers and other more familiar waders on Cley Marsh where we were made welcome by the resident warden Billy Bishop.

Cley Marsh was then, and still is, a reserve owned by the Norfolk Naturalists' Trust, although I seem to remember the organisation having a longer-winded name in those days. If the Solitary Sandpiper took refuge there it would have been safe from the attentions of Arnold, as shooting was forbidden and, as I recall, only members of the society were allowed there, at the discretion of Mr Bishop of course. Much of the reserve consisted of reed-beds, which hid several brackish pools, some of which were accessible, others not, and these represented a wonderful refuge for a variety of species at all seasons. It is many years since I last visited Cley but I understand that in many ways the reserve area has not changed a great deal, except that it has been made much more accessible.

Another feature of birding at Cley which has not changed a lot is the famous East Bank, probably the best known vantage point in Norfolk, affording views of the marshes on both sides and also providing access to the beach. The marsh to the east of the East Bank has also altered little by all accounts, but few present-day birdwatchers are probably aware that Arnold's Marsh, as it is called still, was once the property of E.C. Arnold, the very same gentle-

12. *Familiar waders on Cley Marsh*

man who was gunning for that Solitary Sandpiper back in 1947. Anyway it was more easily surveyed from the East Bank than the reserve was and consequently most of the interesting waders were found on Arnold's Marsh with less effort. There was also Salthouse Marsh, still further east, which one could walk out onto by means of a series of rusty steel planks, probably relics of wartime activity.

Then there was the beach road which led to the old coastguard station, its derelict buildings and a few scrubby bushes, tangles of barbed wire and fence posts ideal for freshly arrived migrants. It was here that I saw my first Hoopoe on the very last day of my holiday. The excitement of that event no doubt to some extent cancelled out the feeling of dread that had overcome me at the prospect of starting my new institutionalised life at Gresham's.

But, before one came to the beach area, there was in those days an encampment of what we referred to as 'displaced persons', refugees from Eastern Europe who lived in a series of Nissen huts outside which they had made elaborate little crests and insignia representing their home towns in Latvia, Estonia or wherever, composed of pebbles and shells laid out on the ground. I mention this for no other reason than to explain that for the first time I became aware that, even though the war was over, for many people their troubles were far from being forgotten. Although I didn't understand, it was I suppose the first sign of some sort of social awareness.

Blakeney Point could be reached from here by means of a long walk along the shingle which could be rewarding, especially if an east wind was blowing – Bluethroats, Yellow-browed Warblers and other mouth-watering rarities could be found if one was prepared to flog through the *Suaeda* bushes which lay just inland from the shingle ridge, though we never managed to find more than a few Redstarts and Willow Warblers on the couple of occasions we attempted it! However, the mud-flats on the landward side were usually teeming with waders and other shore birds and so even if the hoped for rarities didn't materialise, there were always compensations.

To return for a moment to the personalities we encountered on that first visit to Cley: Arnold and Higgins were undoubtedly the ones that made the greatest impression on me, probably because of the way everyone reacted to mention of their names. The two always went together and, as I now realise, Higgins was Arnold's companion cum 'hit man' to borrow a present-day expression which would have been meaningless in those days. Arnold must have been pretty ancient by then, and probably unable to handle a gun on his own. He was quite a talkative old boy and, inevitably, reminisced about the Eversmann's (Arctic) Warbler he had shot at Blakeney in 1922 and other rarities of the past. Of course he was of a generation which thought no more of shooting a rare bird than I

The Juvenile Spreads his Wings

13. My first Hoopoe

did about bottling a Bath White but, blind to the similarity, I viewed him as a kind of ogre, and Higgins his henchman was of the same ilk!

Next door to the George lived another old gentleman named Borrer whose passion was also for stuffed birds. He had a fine collection of them in a kind of private museum, which Father and I were privileged one day to be invited to see. I really don't recall the specimens but do remember his parting remarks when he learnt that we came from Devon.

'Really must get down to visit you some day,' he said. 'I'd like to come down in spring and shoot a family of Ravens on the nest'!

Imagine the horror that such a remark would evoke these days; we were pretty amazed then but old Borrer just did not see anything wrong in the idea! He and Arnold shared the opinion that 'what's hit's history – what's missed's mystery', and would not be likely to change, so he was another one from whom information of rarities had to be withheld. We hear a lot nowadays about 'suppression' of the whereabouts of rare birds, but the phenomenon is not a new one, and brought about for much the same reason – in the interests of the bird.

Not all the other characters were baddies fortunately, and Mr (or was it Major?) Daukes was one of the few who commanded our respect. I cannot visualise him too well, but I seem to recall a quiet schoolmasterish man, obviously an experienced ornithologist, and probably the holder of the secret of the Solitary Sandpiper. D.D. Harber impressed me in a different way altogether – I didn't take to him at all! He was an irritable sort of chap of indeterminate age (children are notoriously unable to tell the age of adults over the age of about sixteen, even if they can tell the difference between a Dunlin and a Sanderling at a glance), given to outbursts of volubleness about the incompetence of other observers, especially precocious schoolboys (or so it seemed to me). Since I didn't have a very high opinion of his expertise either, I never felt comfortable in his presence, though Father apparently got along with him very well. He had a military-style moustache, greying hair brushed back from his forehead and an abrupt manner which I probably interpreted quite wrongly, though I found out later that he rubbed up many other bird-watchers the wrong way over the years.

Also living in the village were 'the Meiklejohns', comprising the noted ornithologist and humorist Maury Meiklejohn, and at least two ladies, one of whom was probably his wife, the other perhaps a sister or a friend – it made no odds to me; but the way everyone seemed to defer to the Meiklejohns impressed itself on me, especially when I learnt that Maury had been a pupil at Gresham's. They were informal people and I remember being invited to lunch at their house and being amazed to find they didn't use plates, but ate straight off a huge wooden table!

14. 'Baggers'

Inexorably, the day finally came when I had to say goodbye to my parents and start my new life at Gresham's. It was not half as bad as I had feared and I soon made friends among the other new boys, though none of them seemed to share my passion for wildlife. I was boarded at the Old School House, situated in the town of Holt itself and about half a mile from the rest of the school buildings. To get to the main school each day it was necessary to own a bicycle and, although I had never possessed one before and had had to take a crash course in cycling earlier that summer, I soon found that my new bike opened up great possibilities.

Very near to the town was an area of heathy woodland known as

Holt Lowes, where I was to spend much of my spare time cycling and walking in pursuit of birds and butterflies during the years ahead. On Sundays after lunch I could hop on my bike and be at Cley within half an hour or so if there were no diversions along the way, though there often were since the whole area was rich in wildlife of all sorts. Gresham's offered much greater freedom than the previous schools I had attended, and I took full advantage of this.

There was also 'Baggers', our geography master, known away from the school as R. P. (Dick) Bagnall-Oakeley, later to achieve fame as a TV personality in East Anglia, but television was not to be a feature of any of our lives until a lot later. Baggers was not only a good all-round naturalist, but an excellent photographer, sportsman, raconteur and most importantly a friend to any of his pupils with like interests. Not surprisingly he was very popular among the few of us who were bird-watchers, and clearly enjoyed the company of youngsters who shared his enthusiasms. He would sometimes invite one or more of us to accompany him on a trip to the Norfolk Broads, Holkham Park, Scolt Head or some other good spot outside our normal cycling range. These visits were always made in his rakish pre-war Aston Martin sports car, in which we bumped (at what seemed like an enormous speed) along country roads, dirt tracks and sometimes open heathland in pursuit of the latest reported rarity. Baggers was always the recipient of any 'hot' news, and gladly let us in on it if possible, though of course it had to be outside school hours. Often we would stop off somewhere for tea and cakes as well, which made the trip even more enjoyable. Other boys among the select little circle which went on these trips included Hugh Ennion, son of the well-known naturalist and artist, and Jeremy Spooner, a childhood friend of my early days at Newton Ferrers who once more became a close companion at Gresham's. Naturally, I suppose, the other boys resented the favoured treatment we received through our association with Baggers and inevitably we were scornfully referred to as 'baglets', though I don't remember ever being inconvenienced or disturbed by the appellation.

Gresham's was not all cream cakes and bird-watching though; I found myself involved in many other activities besides the routine lessons and games, at which I acquitted myself moderately. Among my other interests were art, poetry and jazz, the latter at first only manifesting itself by collecting records, but which soon became an obsession equalled only by my love for wildlife.

My first exposure to jazz had come a couple of years previously at Newton Ferrers during one school holiday. I had attended one of the occasional village film-shows with my parents as a special treat. The feature film I do not remember at all, but before that started a film called 'Jam Session' was shown, which featured various jazz and swing bands of that era, among them I think Louis Armstrong, but certainly a band called Woody Herman's Woodchoppers who were filmed playing in a woodland clearing. Why this image particularly should have fixed itself in my memory I don't know, but I really enjoyed the foot-tapping music, and was very worried when after the show my parents complained bitterly about the awful film with its dreadful music which we had been obliged to sit through while waiting for the main feature – which clearly had made no impression on me. I began to think that if I had enjoyed the jazz, which was evidently not decent music, I must have depraved taste or worse, so I kept quiet about it. In fact I think my parents and many people like them in those days suffered from a kind of cultural snobbery that has become comparatively rare in recent years, but it was perhaps on account of the 'unspeakable' nature of jazz at home that I found it so attractive! Anyway I found at Gresham's that several of my friends enjoyed listening to swing and boogie-woogie which were all the rage in the forties, and I gladly joined the little coterie of enthusiasts and started collecting records. I did not possess a gramophone, but some of my friends did, and between us we were able to assemble an odd assortment of records that would have given a jazz purist a coronary! But we were not then hidebound by conventions of taste and authenticity which I later found to be quite a handicap even during my playing days.

During my second year at Gresham's it became evident that I had torn a cartilage in my right knee. It was decided that the only solution to this problem was to have it removed – not in those days the simple operation it has since become, but involving many weeks of subsequent inactivity, crutches and electrotherapy. This severely curtailed my sporting and bird-watching activities, but gave me lots of time to listen surreptitiously to radio jazz programmes and brood unhappily about not being able to get to my beloved Cley at weekends during the following summer. It did, however, give me the opportunity to work on my entry for the school natural history prize, which I won the following year with a paper entitled 'The Birds of the Yealm Estuary South Devon'. Following this success great things were predicted of me in the future, and it was generally expected that I would follow in my father's footsteps and become a research biologist. I suppose I went along with this idea too, or at least gave no indication that I had any other thoughts about my future career – the truth is I really had no interest in the future, being only concerned with the present, in company with most of my contemporaries I suspect.

Meanwhile I was beginning to assemble quite a collection of jazz records and, being no longer able to conceal my secret vice from my parents, decided to make a clean breast of it. They were simply flabbergasted – they could not believe that their son, such a quiet boy with respectable hobbies like bird-watching and butterfly collecting, could also be interested in such a decadent form of music. But to their credit they did not attempt to nip the nasty habit in the bud, although making their disapproval of my poor taste very obvious. They even helped me to buy my very own record-player, hoping perhaps that one day I would graduate to real music, though attempts to get me to learn the piano in the past had not met with much success.

Art had also always been a minor interest since my earliest school-days when I had shown some promise at painting and drawing. For a while I had had lessons from a practising painter living in Newton Ferrers called Henry Cogle. Cogle's landscape oils and water-colours had been quite popular among the village residents and my parents had purchased several of these. He taught me the rudiments of composition and perspective drawing, at which I had become fairly competent, but it was in the area of bird portraiture that all my artistic abilities were channelled at Gresham's; not only in the weekly art class, but in the margins of notebooks on every imaginable subject would appear thumbnail sketches of Greenshanks, Avocets and my other favourite waders. Baggers wrote my geography report that year as follows: 'If he paid as much attention to his school work as he does to the drawings in the margins of his notebooks he would do well.'

The autumn of 1948 was almost as memorable as the previous one – largely I suspect because of the discovery of a Pectoral Sandpiper at Salthouse. In those days this was still regarded as an extreme rarity and because of its extreme tameness the bird was easily photographed by Baggers and Philip Wayre, who set up a plate camera on a tripod in the middle of the bird's favourite feeding area and just waited for it to come within range. This it did quite happily, even on some occasions coming too close to be in focus! The Pectoral Sandpiper's tameness became its eventual downfall, because legend has it that old Borrer slipped out one early morning before any other watchers were about and caught it for his collection using a butterfly net!

15. *An early rarity – Pectoral Sandpiper*

The Juvenile Spreads his Wings

16. *A heavy crop of Waxwings*

Later that autumn I saw my first Snow Buntings and Shorelarks, and the following February had the overwhelming experience of discovering a party of about forty Waxwings, sitting in a bare tree beside the road as I was cycling towards Blakeney. Those beautiful birds were contentedly puffing out their plumage and preening themselves after gorging on hawthorn berries from a nearby hedge. They looked like overripe fruit themselves and I will never forget how thrilled I was to see my first-ever Waxwings.

My memories of those days at Gresham's are rather confused in my mind today, and I suspect that I have slipped up somewhat in the sequence of events, but somewhere about now I should introduce another of the major influences on my young life, a man whose enthusiasm for birds and his encouragement of young bird-watchers equalled, perhaps even surpassed, that of Dick Bagnall-Oakeley. Those of my readers to have visited Cley at all during the period from the late forties through until the mid-seventies could hardly have failed to come across the genial figure, clad in beret and leather jacket, who patrolled the area of the East Bank and its environs – bird-watching or socialising it was all one and the same to him. Richard Richardson set up and ran single-handed the Cley Bird Observatory for the duration of its too short existence and it was during my years at Gresham's that he arrived there, not long after his release from active service in the Far East during the war. He at once made a huge impact on me, and all the other youngsters he met.

Anyway, I think it must have been in the autumn of 1949 that Richard first arrived at Cley. He set up his observatory in the derelict buildings of the old coastguard look-out, and utilising the sheltered south wall of a former gun-emplacement constructed a very effective Heligoland trap where he was soon catching and ringing a stream of migrants. We were only too glad to have the chance of helping out in this project, picking up at the same time all kinds of tips regarding bird identification, especially learning how to tell the various wader calls and even, under his expert tuition, learning how to imitate them as well. I say 'we' because there were several other young hopefuls always hanging around Richard, among them Peter Clarke, now warden of the Holme Observatory further along the coast, Graham Byford and Jeremy Spooner, both fellow-pupils at Gresham's. Other enthusiasts, among them M. J. Rogers and A. R. Mead-Briggs, were regular weekenders from as far away as London and Birmingham, for the fame of Cley was already spreading far and wide!

Richard was also working at that time on his illustrations for *Collins' Pocket Guide to British Birds*, the first ever field-guide to appear in Britain. His paintings were superb: they would stand comparison with those of any modern guide, but he was badly let down in the reproduction process and, apart from pen and ink work, he rarely allowed his paintings to appear in print after that, preferring to make his living from originals or commissioned portraits, which he often sold for prices well below their value. But although obviously a complex character his needs were modest; he never seemed to want the fortune that could have been his in a commercial sense. Fame came inevitably, whether he wanted it or not. How I envied his ability to draw birds so effortlessly – he seemed able to convey the 'jizz' of a bird with only a few lines of pen or pencil – a talent that few are lucky enough to possess. I knew in my heart that I would never be able to emulate it, though God knows I tried!

In the lengthy periods when I was either unable to get to Cley or less inclined to do so than in previous years, other matters were

17. Richard Richardson

increasingly claiming my attention. My jazz aspirations had taken the first step in the direction of performance with the purchase of a second-hand trumpet, beautifully ornamented in silver and gold lacquer, which I was attempting to learn to master without the benefit of formal tuition. My housemaster, no doubt with keeping the peace in mind, insisted on my practising in the vegetable garden, well away from the school buildings. Other budding instrumentalists were also practising in other houses, the gym or cricket pavilion, with every intention eventually of forming a band, though our first aim was for each of us to get some individual proficiency before coming together for a joint effort – still only a far-off projection for the future.

I was also involved in a tortuous emotional relationship with another boy – not an uncommon situation at boarding school, but in my näiveté I thought that, like jazz, this was another dreadful flaw in my make-up that I had no control over. But worse than jazz, I could never admit the fact to parents or teachers, although in retrospect I realise it must have been patently obvious to the latter, even though there was something of a tacit convention among pupils and staff that such liaisons were never acknowledged for fear that they might through being revealed even become acceptable. So the various attachments I noticed among other pupils at Gresham's were probably just as emotionally disturbing to those concerned, but seemed as nothing when compared to my own particular tangle. I have no wish to dwell on this aspect of my life at school, but it may help the reader to understand how a sensitive creature, which I definitely was in those days, could through emotional disturbance become introverted to the extent of burying myself in all-absorbing practices like teaching myself the trumpet, to the exclusion of practically everything else.

No mention has been made of my school work at Gresham's which during the first few years showed promise, resulting in my achieving better than average results in the School Certificate Exams (in those days the equivalent of GCE 'O' levels). But when the time came to decide on specialising for the newly introduced 'A' levels I had difficulty in making up my mind, torn between my obvious talent in the field of biology (though not in other sciences) and my secret wish to dabble in what were loosely referred to as 'Arts' subjects. I read a lot, including novels and poetry of a leftish nature, and together with several of my contemporaries considered myself a bit of an intellectual. My jazz interest helped to foster this notion and, having always been a bit of a rebel at heart, I really longed to strike out and do the opposite of what my parents and teachers all expected me to do – become a biologist like my father. Of course when it came to making the decision I didn't have the courage to speak out and limply followed the path that had been prepared for me.

The visiting careers adviser was not much help either. When I had my obligatory interview with that gentleman, I decided to enquire about the possibilities of a career related to my main interest, natural history. He smiled pityingly. 'Young man I think you should understand right away that jobs of the sort you are asking about do not exist. My advice to you,' he continued, 'is take up banking or accountancy, then you'll be able to earn enough money to indulge in your natural history interests in your spare time.' And that was the end of the interview as far as I was concerned. I could think of nothing I wanted to do less than take up banking or accountancy and, since my only ideas for a career did not exist, it seemed we were wasting each other's time.

So at the age of sixteen, with an intense love of wildlife, a fascination for jazz music and in an emotional turmoil concerning my sexuality, I took the only way out of my dilemma and followed the advice of my teachers, which was to concentrate on the sciences of botany and zoology, do sufficient of the other sciences to get me a place at Oxford, where Father had already made enquiries about a place at Exeter College, and hope for the best!

Fully recovered from my knee operation I was also involved quite heavily in sport, especially hockey at which I excelled as a goalkeeper, being a member of the school first team at the age of sixteen; I was also a fair rugby fullback and wicket-keeper during the cricket season, although I never represented the school at either. So in one way or another I found my time fully occupied during my last two years at Gresham's – the one thing I did not find a lot of enthusiasm for was work though, and even my biology subjects suffered.

As they had at St Faith's, school holidays came as a blessed relief and in September 1950 and April 1951 Father and I visited the island of Lundy in the Bristol Channel. There I was able to immerse myself wholly in the bird-observatory routine of trapping, ringing and watching birds from dawn to dusk. It was like genuine therapy after the fraught weeks of school and I was able to relax and enjoy life, even to the extent of unburdening some of my worries to Father. I told him everything – well almost everything, and quite naturally he was worried about my future as well as my present state of mind to the extent of arranging for me to visit a psychiatrist.

A former colleague of Father's, a Cambridge don who had also had problems with his sons, was able to recommend a specialist, so I was packed off for a course of treatment at Cambridge during the spring holiday. I only had one session with the psychiatrist, but long enough to take a dislike to him. And then fate stepped in by giving me a dose of mumps and I was packed off to the Isolation Hospital. During the course of my convalescence I fell rather heavily for one of the nurses, thus releasing me from my greatest fear – that I was condemned to a life of homosexuality! Of course I had never been in contact with a girl before in my life, at least not in a way in which it had been likely I would feel sexual arousal, so it was not surprising that my impulses in that direction had until then only been engendered by other boys. Gresham's has now a number of girl pupils as well as boys, surely a much healthier situation than in my day. Not that I believe Gresham's was different from any other public school in that respect – just the one I had personal experience of.

So my final year at school was a lot happier than the previous one had been – I had fewer emotional hang-ups, I had dropped physics and chemistry from my 'A' level plans and substituted English literature, which I not only enjoyed but fancied I had a good chance of success at although it did not fit too well into the scheme of things, my other main subject being biology. However, I was content with this compromise and in my spare time kept on with my trumpet practice!

The idea of a school jazz band had been officially accepted on one condition – that we gave a concert at the end of term! So with a bunch of ill-assorted instrumentalists, in the most unlikely combination and with the worst possible clash of styles, we sought to put a programme together. The resulting mayhem, which took place in the main assembly hall known as Big School, was given a whole paragraph in the school magazine as follows:

> Playing many well known jazz tunes a spirited quintet performed to an audience of over a hundred boys and several staff. This was almost certainly the first recital of its kind in the history of the School.

If the standard of playing did not improve in the future I sincerely hope it was the last recital of its kind, but I intentionally lost touch with life at Gresham's after leaving in the summer of 1952 and it was over twenty years before I had any further contact with the school. My 'A' level results were a miserable failure and I guess I felt I wanted to forget everything about the five years of my life I had spent there. My schooling ended on a sour note.

3. Years of Immaturity

When I left Gresham's at the end of the summer term in 1952 two, or perhaps it was three clouds were hanging heavily over me. The first was that after a poor showing in my 'A' levels I had no chance whatever of getting to any university, and thus was quite unable to see any way of shaking off the second – the threat that in those days hung over every school-leaver not destined to further his education – National Service for two years in one branch or another of Her Majesty's armed forces. Although the war in Korea was just about over, other conflicts seemed likely to crop up; it was an alarming thought that I might be pushed into some military conflict in a far-off land and have to fight for a cause in which I didn't believe and would therefore have no idealistic spur to urge me on, even if I had been an aggressive type. In those days I was quite the opposite with my fuzzy left-wing political notions and love of poetry and jazz!

To make things worse, whether I survived National Service or not, I still hadn't the faintest idea of what I would like to do as a career apart from play in a jazz band! All notions of a nice open-air job connected with my natural history interests had been dispelled long ago by the careers adviser, and it was this uncertainty that formed the third cloud hovering darkly over my future.

My life until then had had little in it to fear apart from harassment by fellow pupils and disapproval of parents and teachers at my inability to live up to expectations. Since I had already experienced plenty of both without much ill effect, such worries no longer bothered me, but the prospect of being seriously injured or even killed in defence of someone else's freedom did not appeal to me one little bit!

So when I failed to make the grade in my first medical examination it was quite a relief, although worrying that they did not consider me fit. But it was only a breathing space for after a chest X-ray and a further examination, in which they could find nothing the matter with me, I was pronounced ready for service and graded A1.

18. Migrant watching in Norway

The delay caused by all this did give me the chance, together with Father, of taking part in a bird migration study in southern Norway before my eventual call-up, and for a while I forgot my worries about impending service, and even my aspirations in the world of jazz!

The Norway trip was my first ever journey abroad, and I remember being enchanted by everything Norway had to offer – from the picturesque but practical wooden dwellings of the fishing-folk with whom we stayed, to the majestic and spectacular scenery of the fiords and forested hillsides.

Our main task was to rise before dawn, and after a superb breakfast of open sandwiches dressed with various cheeses and fish, washed down with a glass of really fresh milk, we would climb to the top of a hill overlooking the southernmost tip of the Norwegian peninsula from where we counted the waves of migrant birds, mostly thrushes and finches which were flooding south at the onset of autumn. The visible migration eased during the mornings and the rest of the day was free to explore the surrounding countryside. Here I saw my first Temminck's Stints, Nutcrackers and various new woodpeckers. I also came to sample the edible qualities of Cormorants and gulls among other local delicacies – though only because they were served to us in the normal course of events, being part of the regular diet in the fishing village where we stayed. I also remember vividly the sensible arrangements in the outside lavatory, which was constructed over the foreshore and regularly flushed by the tide!

We travelled to and from Norway by steamer from Newcastle,

Years of Immaturity

19. *Temminck's Stints*

and our return journey was into the teeth of a gale, taking about a day longer than scheduled. Being prone to seasickness as I am it was another memorable experience, but it did not stop me from gorging myself on the delectable smorgasbord provided – indeed I was able to consume more than usual, since I usually jettisoned a meal soon after eating it and rapidly became hungry again! One distraction from all this was the appearance on board of a Wren, which had presumably stowed away while the ship was docked in Stavanger – whether it survived the journey as we eventually did we were unable to ascertain.

Back in Britain once again there were no further hindrances to the inevitable process of conscription; when the time came for me to choose which branch of the services I wished to join I opted to sign on as a regular for three years in the Royal Army Education Corps. My motives for taking this option were partly because several of my school contemporaries had already done the same and seemed to be surviving, but mainly I suspect because it seemed

20. *'I also came to sample the edible qualities of Cormorants and gulls'*

21. *North Sea stowaway*

an easy way out. The prospect of an extra year in the army was not too daunting, especially as I had no concrete plans for the future. After a six-month's training course I would become a sergeant instructor. I would earn far more than ever possible for a National Service conscript, and on top of this I had a good chance of being posted abroad and little chance of having to take part in any hostilities. What better way could I find to take me through the dreaded but obligatory period of National Service?

I left my trumpet and my binoculars behind during my basic training, which took place at the depot of the Middlesex Regiment (The Diehards) at Mill Hill in North London. I do not recall too much about those first few weeks in the army, except being agreeably surprised that it was not unlike life at boarding school all over again! The main differences were that most of my fellows did not come from sheltered middle-class backgrounds, many of them never having been away from home before in their lives, and the language, which being largely in the street vernacular of London I found pretty hard to understand. What my barrack-room mates thought of me and my posh accent I have no idea, but they soon found out that I could keep my head well above water in the struggle to complete basic training – my years at boarding school really came in handy in that respect and I had little difficulty in surviving those early weeks. I soon learned to disguise my public school background by slipping into the army style of speech – horrible though it had sounded to me on first hearing it I was soon 'effing and blinding' as well as anyone in the barrack-room, and although it probably sounded pretty affected it served to make me 'one of the lads'.

After completing six weeks' basic training I was sent to the RAEC instructors school at Beaconsfield in Buckinghamshire. From there I was able to obtain the occasional weekend pass which enabled me to hitch-hike up to London to hear one of the revivalist jazz-bands currently playing in the city, like Ken Colyer's New Orleans Jazzmen or my idol Humphrey Lyttelton at the 100 Club in Oxford Street. Those sweaty evenings were the nearest thing to heaven that I could imagine, and I used to return to camp recharged for another week of instruction. Bird-watching and other country pursuits were forgotten during my early days of National Service, and indeed, for the next decade or so were destined to take a back seat.

Actually I dropped out of the education instructor's course after three months – I found I had neither the aptitude nor the patience to put up with the ignorance of my fellow soldiers, many of whom I discovered were not only stupid but illiterate; it was the instruc-

22. *Me and my army mates – that's me on the right*

tion of illiterates, I learned, that was the main task of the sergeant instructor. I decided that since I had to complete my service as a soldier I might as well do the thing in style, and rashly put my name down for a parachute course – a reaction I suppose to the weeks of classroom study I had been stuck with at Beaconsfield. Luckily, I suspect my history of knee-cartilage trouble did not allow me to become a paratrooper, so I never learned whether I had the nerve to jump out of a flying aircraft or not. My second choice had been the Devonshire Regiment – after all they were stationed in Kenya which seemed to be an exciting place to go, and as a Devon man I felt this might give me the required posting. As it turned out I was transferred to the 'Glorious' Gloucesters who had just returned home in triumph, but in a decimated state from the battlefields of Korea. After that demanding tour of duty it seemed the Gloucesters were due for a home posting, and so it was that 22836280 Private Hunt found himself not battling Mau-Mau guerrillas in East Africa, but demonstrating how to dig trenches that would withstand tactical nuclear weapons on Salisbury Plain! It was a change from the classroom, and nice to see the occasional Buzzard soaring overhead or glimpse a fox on the skyline, but it was still far removed from what I had hoped the transfer from Beaconsfield might achieve.

One day our battalion were put on standby to fly to Jamaica to take the place of troops who had been rushed from there to quell an uprising in Guiana. We had yellow-fever injections and were fully prepared for the move, but it never happened and instead our unit moved from Warminster to Barnard Castle, in the bleak northern fells of the Pennines – a far cry from the hoped-for warmth of the Caribbean! Life at Barnard Castle was made almost unbearable by the attentions of Sergeant Major 'Shitter' Courtenay, the sort of non-commissioned officer depicted in TV programmes or 'Carry on' films about whom one can have a good laugh in retrospect, but when encountered in real life is far from a laughing matter! Shitter seemed to have it in for me from the start – he was not fooled by my attempted West Country accent, clearly seeing me as an officer-type who he could actually get the better of. Regularly during my training periods I had been encouraged because of

23. *Birdwatching army style*

my public school background to apply for a commission, a temptation to which until then I had not succumbed, so perhaps just to get away from Shitter, but also because I saw it as some kind of challenge, I applied to go for a selection board – the first step towards Officer Training School. Shitter seemed quite pleased about this – I suppose he looked forward to humiliating me if I came back having failed!

Before the selection board came up I applied for, and got to my surprise, a long weekend's leave which I decided to use in a mad dash by train to London and back, in order to attend one of the all-night jazz sessions which used to take place on Saturday nights at either the Colyer Club or somewhere in Soho. All I remember of that weekend is waking up in the Military Hospital at Woolwich suffering from a rare kind of dysentery. After a few days convalescence I returned to my unit to meet a beaming Shitter.

'You're on a charge,' he roared, 'absent without leave'. He was absolutely delighted. It seems that the hospital had not informed the unit of my illness, so it was a bitter blow for Shitter to find my hospital discharge papers in order.

I passed the selection board with ease – my examining officer turned out to be a keen naturalist and we chatted throughout my interview about birds and badgers in Devon and Norfolk. So I was able to bid Shitter a fond farewell and leave the joys of Barnard Castle behind me when I transferred to Officer Cadet School at Eaton Hall in Cheshire. Here for the first time during my army career I actually found myself among kindred spirits, most of whom were also from public schools, some of them as exclusive as Eton and Harrow. Actually I found the ex-Etonians to be not at all the snobbish types one is led to expect. One of them, Alex McEwen, proved to be a passable guitar-player with a knowledge of jazz and folk music and we had many 'sessions' on evenings when we were confined to quarters. However, we were often able to escape into nearby Chester where we soon had firsthand experience of every bar, especially the Quaintways Jazz Club. There I remember hearing the legendary Christie Brothers Stompers as well as the inevitable Humphrey Lyttelton, who I reminded myself had not only been an Etonian but also was an ex-officer. I seemed to be going the right way about becoming a famous jazz-trumpeter, even if I had still never played in a proper band! After the training course, which also included a lot of hard physical effort, more days in the classroom, field experience in Snowdonia and battle camp on Dartmoor, I passed out as an officer if not truly a gent!

To my delight I was commissioned to the Devonshire Regiment, still serving in Kenya – at last my dream of a foreign posting seemed to be coming true. Imagine my disappointment though, to find that because I only had about six months of my National Service to complete, I was to be sent to the regimental depot at Exeter! And so I whiled away my remaining weeks in the army as assistant adjutant, my most important job being to arrange cricket fixtures with other units and local teams. Being the only officer young enough to play as well, I also found myself having to be captain, and answerable for all ensuing defeats! Of course there was

24. *A youthful artistic endeavour*

always the occasional visit to the firing-range to conduct, and periodic duties as Officer of the Guard, but my military career came to a halt as inconspicuously as it had started and I have done well to string it out to more than half-a-dozen paragraphs.

Having finished my National Service I was again faced with the problem of what to do with my life, but unbeknown to me this had already been decided for me, at least in the short term. My Mother, while I had been risking my life for Queen and country in far-off Exeter, had collected together a 'portfolio' of my artistic output dating back to my early efforts at St Faith's. It included drawings of birds, landscapes, abstracts and other ghastly juvenilia which by then I was thoroughly ashamed of. She had then taken this collection of garbage to the principal of a London art school who had agreed to accept me as a pupil, subject to my being able to pass an interview. I would never have dared to show what passed as my 'work' of the past years to an art school but Mother had done so and I had been accepted on the strength of it, so it was not difficult for me to convince the head of the Central School of Arts and Crafts that I was hell bent on a career in commercial design – I had all the motivation I needed. Through contacts I would surely make as an art student it would be only a matter of time before I graduated as a fully-fledged jazz-trumpeter. What better way could there possibly be for me to get on the scene? My idols Humphrey Lyttleton and Wally Fawkes and others were all ex-art students or practising artists so, although she didn't know it, by helping me get to art school my proud mother had in fact aided me in my first step towards fame and perhaps fortune (though of course the latter was unimportant in those days) as a jazzman!

That in fact is about the only fantasy in my life that actually came true, and I often wonder what I might be doing now if my mother had not so lovingly touted around those truly awful artistic endeavours of my youth? The fantasy of fame and fortune unhappily did not quite materialise; but the years of touring at home and abroad as a member of various bands, most of them totally obscure, and earning peanuts, are a fond memory and my life would certainly not be what it is now if I did not have those years to look back on.

The details of those days at art school, of my life in London and Germany as a musician, are best glossed over not because they are uninteresting, far from it, but because they are not strictly within

25. *Refreshments in Andorra*

the scope of this book, though in the pages that follow I will outline the basic sequence of events as they happened to me for the sake of continuity – with a little anecdotal material to stop it becoming too boring for me as well as for you. If you are really only interested in the 'birdman' bits and don't care how the art student finally became a jazzman or how he evolved into the birdman, I suggest you skip the rest of this chapter and the next, because although they are quite amusing you won't like them very much.

My love of wildlife remained dormant through most of the period 1954-64 though there were occasions when it had the chance to show itself. I vividly recall the Easter vacation of 1955 when Kit Kitson, an art school friend and I, with £10 apiece took off for a hitch-hiking holiday to Andorra in the Pyrenees. How we did it on so little money I cannot now understand, but I have photographs that clearly show we got there, and an old passport which indicates the pitiful state of my finances at the time. My main recollections are of the Camberwell Beauties, Queen of Spain Fritillaries and other exotic butterflies which were coming out of hibernation and sunning themselves by mountain streams still embanked with snow and ice, and omelette breakfasts washed down with black coffee, with sherry from the wood at the equivalent of sixpence a glass! During our short stay in Andorra Kit and I found ourselves cut off by a snowfall which blocked the pass leading back into France. Being young and foolish we decided that we would walk out by an alternative route. We succeeded, but had to spend one night in a sanctuary for lepers, and during the course of our walk my hands became so badly sunburnt that for the next day or so I had to walk like a somnambulist with my hands held out horizontally in front of me, to alleviate the pain caused by the fluid in the blisters on the back of my hands trying to respond to the forces of gravity.

Undaunted by this experience I returned to Spain in July of the

26. Bee-eaters overhead

same year, joining some other art students to share a villa in Andalucia for a few weeks. We were the only foreigners to be staying in the little hill village of Mijas, and even in the nearby towns of Fuengirola and Torremolinos we met very few other British visitors! I have never been back to Andalucia since the massive surge of popularity of the area as a tourist venue – I wish my memory of the area to remain unsullied – a land of peaceful beauty with the song of peasants working in the fields and the incessant calling of Bee-eaters overhead the dominant sounds. Other birds I encountered for the first time on that holiday were Black Wheatears, Rufous Bush-chats, Sardinian Warblers and many other typically Mediterranean species that whetted my appetite for more exotic bird-watching than I had previously experienced, but it was to be a long while before I would have another chance to enjoy such pleasures.

27. *The art school image*

Years of Immaturity

My first year at the Central School had been a great time of self-discovery. I had a completely new image now, and few of my army pals would have recognised the skinny creature with an unruly mop of hair, in tight black jeans and hairy jumper, which I had turned into. The main objective of each successive generation of art students is to be the opposite in every way of the previous one and to this end we went to absurd lengths to shock our predecessors. Today our appearance as it was then would verge on the fairly conservative, but back in the fifties we looked like creatures from another planet! My poor parents would have been aghast at my new image, but whenever I went back to Devon, which was very seldom, I contrived to return as much as possible to the conventional hair-style I had forsaken and donned sports jacket and baggy trousers in order to keep the peace. It amuses me now to think that about five years later our art school image of the mid-fifties was adopted and adapted by pop groups all over the world, who thought it had been originated by the Beatles!

Another eventful thing that happened to me at the Central was the discovery that I liked the company of girls, and what's more they seemed to like me too – something that during all those years at boarding school and in the army I had only considerd as an abstract notion that bore no relation to reality. I wasted much of my first year in dalliance with a variety of females, mostly from art school, but occasionally ones I met at jazz-clubs, and in that time I packed in much of the experience I had failed to obtain in the previous twenty years to the extent that my main ambition of becoming a jazz-trumpeter was temporarily shelved.

In my second year at the Central I got to hear about a band that had been formed at the Slade School called Mike Tyzack's Pagan Jazzmen – they played at the weekly 'college hop', and were sadly lacking in instrumentation, although to my disappointment I found that the trumpet was played by Mike himself, so there was no opportunity for me there. They did need a trombonist, however, so without giving the matter a second thought I traded my shining silver and gold-plated trumpet for a beat up old trombone and within a fortnight had mastered the instrument well enough to be able to rasp and grunt away in the approved style of New Orleans, which also happened to be the easiest style to get away with without previous experience of the instrument! Our playing engagements were mostly art school hops and parties but, after a year or so of this, and having systematically neglected my artistic studies, I felt ready to embark on my full-time musical career. Jazz had still not caught on in the pop world as it was destined to do a year or so later and most bands were amateur in status apart from a handful of big names. There was a demand for bands in Germany, however, and when I heard that a hot trombonist was needed for a six-month tour of German night-clubs I went for an audition; and I got the

Years of Immaturity

28. *A hot night in the Boheme Club*

job. (I should have been surprised, but in my naïveté did not appreciate that available trombonists were in short supply and time was running out!)

So in October 1956 I found myself a fully paid up member of the Musicians' Union, playing with Eggy Ley's Jazzmen in the Boheme Night Club in Cologne. Our earnings were the equivalent in Deutschmarks of about £65 per month – a fortune by the standards I had been used to as a student in London making do on £6 per week. What's more I was playing jazz every night of the week from about 9.00 p.m. until 3.00 or 4.00 a.m. – the fanatic's idea of heaven! We slept all morning, had breakfast at lunchtime and strolled the streets of the city, which was being handsomely rebuilt after the devastations of war. There was little or no opportunity to go bird-watching, but I was so immersed in leading the life of a successful jazzman that such things were far from my mind. Drink was readily available in the club and each night we would start on the beer before commencing our set and continue throughout the night. I have little idea what our nightly consumption must have been but it was certainly more than was good for us; luckily I never developed a taste for harder liquor or drugs, which have terminated the lives of several of my associates from those years, some of them dying even before the age of thirty!

It was during our first month's engagement in Cologne that I noticed among the youngsters that came to dance nightly at the club a girl with alluring eyes and a generous smile, her long dark hair in pigtails. Getting to talk to her I found that she spoke excellent English, including some of the choicest expletives (without a

Years of Immaturity

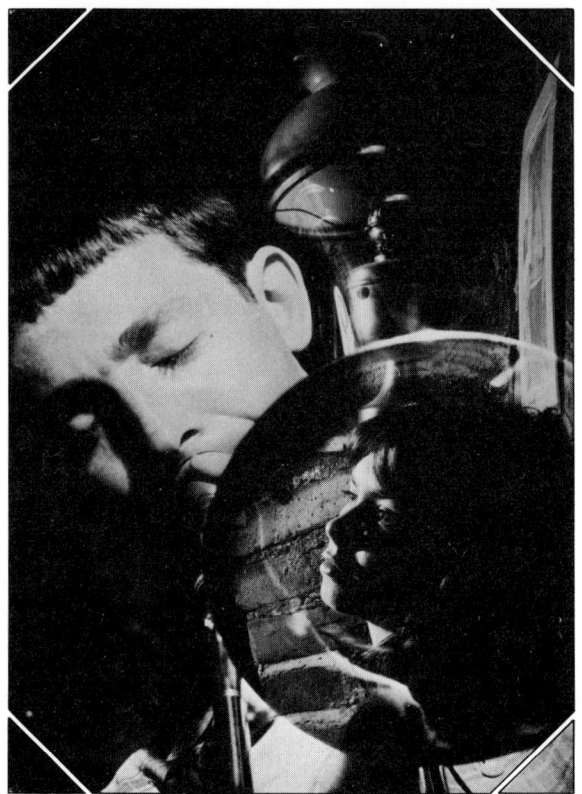

29. *My two early loves*

clue as to their implications). The attraction proved to be mutual, and Marianne soon became my travelling companion as we moved about Germany. We played in Wuppertal, Duisburg, Aachen, Düsseldorf and Cologne again before returning to England for a rest. It was during that time, in the spring of 1957, that Marianne and I were married. Although her knowledge of the English language even then was very good, she did occasionally get confused, and when at our Registry Office wedding she solemnly repeated after the clerk 'I take you to be my awful wedded husband', she was not joking! That we are still married nearly thirty years later is a tribute to her extreme tolerance and understanding, for I fear I have fulfilled the role of awful husband only too well. Luckily Marianne does not share my obsession for wildlife and is therefore always at home to 'hold the fort' during my frequent absences on what I am fortunate to be able to regard as business. I would like to thank her here and now for her loyalty through some pretty bleak times, and for her love and care for me and our children which has contributed so much to my being in a position to write these memoirs. Without Marianne's strength to draw on over the years I would surely be telling a very different story, or more likely none at all.

We attempted to settle down as newlyweds in London and, together with Mike Tyzack, who was then becoming known as a painter rather than a jazzman, and Ian McKerrow, a Scots clarinet-player who I had known since student days, we shared a flat in Highgate, North London. While Marianne got a job as a nanny to a well-to-do family, I looked for a 'proper' job for myself. This proved to be not so easy – after all who would employ an ex-art student whose only qualifications were a bunch of 'O' levels in impractical subjects and whose only work experience had been as a jazz trombonist?

After a lot of searching I finally found employment as a junior sales assistant with Alkits, the gents and military outfitters of Cambridge Circus (passing there recently I see the old premises have become a Wendy's Hamburger Joint!). At Alkits, where I guess they had taken me on because of my distinguished military background, I worked in the hosiery department and was required to sell socks, shirts and even shoes to business types. For this I had to wear a suit (luckily I had bought one to get married in), a clean shirt and tie, and polished shoes daily. Amazingly I did quite well as a salesman and, in order to earn the little bonuses that were continually being offered, I would sell people shoes that did not fit and ties that did not suit them as if I had been highly trained for the job. The head of the department was actually quite upset when I handed in my notice with considerable relief when I was offered another contract with Eggy Ley.

Back in Germany again, it was great to be able to play jazz to my heart's content and resume the life which I had so much enjoyed previously. We visited several new towns, among them Hamburg, and this gave Marianne the chance to stay with her family in nearby Stade while I was working at the New Orleans BierBar on the infamous Reeperbahn. Unknown to us at the time another musical group, later to become known worldwide, were also working out their musical apprenticeship in Hamburg – the Beatles. Rock and roll was of course beyond the pale as far as us jazzers were concerned – we were as scornful about that idiom as my parents' generation was about jazz, so we would not have been interested in the Beatles even if we had known about them in those days.

Our new contract also included dates in southern Germany, an area we had not previously visited, and occasionally we would get a couple of free days. I remember walking in the Taunus hills just north of Frankfurt, where Golden Orioles called incessantly from giant leafy oak trees without ever showing themselves, a wetland reserve on the Rhine near Mannheim where I encountered my first Little Bittern. Wherever we went the lonely song of Black Redstarts

30. *Kites over the Rhine*

seemed to haunt me as I returned to lodgings after work each early morning for, in the cities still partly devastated by the effects of war, this was a species that really took advantage of the ruins.

In August 1958 the band had a month without an engagement, so Marianne and I decided to make the most of this and enjoy a kind of belated honeymoon. This started with a cruise from Cologne, and down the Rhine as far as Mainz. As we cruised we passed thousands of dead fish, the result of industrial pollution, but proving no deterrent to scores of Black and a few Red Kites which were hooking them deftly from the river. We then hitched south to Munich from where we headed for Italy. It was on the first leg of this journey that, because lifts were hard to get, I posed as a female, borrowing a scarf from Marianne, wearing it around my head like a girl and standing in the shadow while Marianne did the car-stopping. It worked almost at once; I scrambled quickly for the back seat while Marianne chatted to the driver. It was not a ruse that we decided to repeat, however, because the driver caused me great discomfort by continually addressing remarks to the back-seat passenger in German, which I did not understand too well and in any case could not answer except with the occasional falsetto 'ja' or 'nein'. When the driver stopped at the next filling station we discreetly vanished, and ended up being transported into Munich by a police car as hitch-hiking on the autobahn was not allowed. I doubt if the police would take such a lenient view today but at that time the dangers to drivers from hitch-hikers, or vice-versa, were not fully appreciated, and I would certainly not recommend the practice now except in unusual circumstances.

We passed through Innsbruck and the Brenner Pass, then headed for Trieste, where we decided to catch a train into Yugoslavia. In Rijeka we were unable to find any overnight lodgings and ended up sleeping on a station porter's wooden trolley. I remember being woken up by the monotonous calling of Collared Doves – my very first encounter with the species which in those days had only just been discovered for the first time in Britain!

Although our holiday was not intended to be any kind of wildlife tour, I could not help wanting all the time to explore the countryside and discover the birds and butterflies; this was a side of my character which until then Marianne had not had to contend with. Although she was not uninterested, she preferred to swim, sunbathe or just relax while on holiday, whereas I am always needing to look around the next corner or over the hill in case it might be hiding something more interesting. Neither of us have changed in this respect, and often go our own ways on holiday. On that occasion, though, we started to suspect that Marianne was pregnant, so I did not completely abandon her. I had to content myself with some peripheral nature study, identifying various warblers, buntings and raptors, and gaining my first experience of snorkelling in the Adriatic. We travelled by ship through the islands as far as Split, then sailed to Venice, from where we made our way back to Frankfurt where my next playing engagement started at the Storyville Club.

The Storyville was a prestigious venue, attracting many American servicemen stationed in the area as well as the locals, and so often engaged big names as well as the resident band. I remember playing on the same bill as Stan Getz, Albert Mangelsdorff and other famous jazz players, and occasionally got to have 'jam sessions' with them. Although we were ostensibly a traditional-style outfit, some of us in Eggy's band had leanings to a more modern style and were delighted to have the opportunity to 'jam' with American musicians. On one occasion we even had a session with members of the Modern Jazz Quartet who were touring Europe at the height of their popularity.

4. The Breeding Period

Marianne's pregnancy was confirmed and when the time came she went to stay with her mother at Stade, where our son Nicholas was born in February 1959 while I was playing in Stuttgart. But, as luck would have it, a six-month residency at the Pigalle Club in Hamburg came up and I was able to be with them throughout the summer. It was a bizarre way of life as I had to become a night

31. *White Storks at the nest*

32. Après Stork

commuter! Each evening I would catch the train to Hamburg (it was about an hour's journey) and after our nightly session would be just in time for the first train of the day – about 4.00 a.m. I arrived in Stade soon after dawn and went straight to bed! But it was worth it to be with the family and in the afternoons we would go for walks, with Nicholas in the pram, in the surrounding countryside. Stade was situated in the low-lying land south of the Elbe, the centre of a huge area of fertile agricultural land largely given over to the production of fruit, especially cherries. White Storks were still a common sight nesting on farm chimneys in those days, and the bill-rattling of displaying pairs was a characteristic sound. Avocets were among the various birds nesting in the unclaimed areas of marshland, and the song of Icterine Warblers was often the first sound I heard on getting up each day just in time for lunch!

Father took the opportunity of visiting us in Stade that summer, and in his hired Volkswagen we did some birding together a bit further afield than my afternoon perambulations with Nicholas had previously allowed. I remember being excessively embarrassed because Father insisted on asking every German peasant we encountered whether they knew where to find Great Reed Warblers – had he asked them for Drosselrohrsänger it might not have been so bad – but to enquire in English struck me as ridiculous! Needless to say we never saw or heard one.

An incident dating from that summer which I can only relate from hearsay, because although I was involved I don't recall much of it, took place on Ascension Day, traditionally a public holiday in

The Breeding Period

Germany and called Himmelfährt (Journey to Heaven). Eggy Ley's Jazzmen had been engaged to complete an exhausting schedule on that memorable day, which started with a river cruise from Hamburg, went down stream as far as Itzehoe, where we were to give a lunchtime concert, followed by an overland journey to Kiel for a third and final performance that evening. I should explain that the German tradition on this day is for the male members of each family to leave home for the day with nothing but a bottle of Steinhager (the local hard liquor) and a chunk of bacon fat (to counteract the effects of the booze). How they are supposed ever to get home again is not fully explained. Maybe the fervent wish of every hausfrau is that her spouse will achieve the ascent to heaven! Whatever the true motive at the back of this annual festival, it was evidently not for the benefit of beer-swilling British trombonists and, as soon as we had completed our first rendition of 'Riverboat Shuffle' or some other appropriate piece, I found stone jars of steinhager being thrust at me from all directions. Since tromboning is thirsty work and there seemed to be little or no beer available, I swigged from every bottle that was offered! By midday the effect was such that I was no longer capable of standing up, let alone playing the trombone. So for our lunchtime concert (I am told) I

33. The stork I never saw

compromised by lying down on the stage and going through the motions. My memory of the rest of that day is extremely vague, but my friends in the band seemed to recall it vividly, so I have to rely on what they told me later about the drive that followed and my disgraceful behaviour *en route* for Kiel. Apparently I was far from out cold, because at one point during the journey I spied a Black Stork in a roadside field and requested a stop – when this was refused I seized one of my companions' shoes and threatened to destroy it if I was not allowed to look at the stork. To humour me and avoid further trouble this was done, but the tragedy is that I do not remember the incident at all and until quite recently Black Stork was still on the list of species that I had never knowingly seen!

As our son grew older it became obvious that we could not continue the touring life indefinitely, so we once more decided to try and settle down in England. I would try to make a living as a touring musician in London and if this did not pay the bills, Marianne could always look for a part-time job if needed. The idea was sound enough but work was not as easy to find as it had been a year or two previously. More bands were contesting fewer jobs up and down the country, and only the well-established names were earning a decent living. By spending so much time in Germany, I had inadvertently cut myself off from the 'scene' in London; most of my former friends had either established themselves in the 'big time' bands or given up, disillusioned. 'Mainstream' style jazz was the coming thing we believed, so many of us who were no longer fitting happily into the traditional set-up tried to get in on this predicted new trend before it happened; I found myself playing in all kinds of bands in a wide variety of styles and all the while earning barely enough to keep us. During this time I worked with so many different outfits that I am hard-pressed to remember their names, but here are a few examples: Fat John's Band, Nat Gonella, The Confederates, Doc Crock and the Crackpots, Mick Mulligan's Band with George Melly and the Professors of Jazz who were expected to play clad in mortar-boards and gowns – needless to say the professors only completed a couple of engagements before disbanding! Actually several of the bands I played in had the same basic personnel, but went out under a different leader or in a different uniform each night under a different name – but that's show business!

Meanwhile we had found ourselves a nice basement flat just off Kensington Church Street, with a little garden where young Nicholas could safely play. Unfortunately my earnings as a freelance musician barely paid the rent, so once again it became necessary for Marianne to look for a job. A West End night-club, the Bal Tabarin, had decided to lift the idea of Playboy Bunnies from America, and was looking for suitable girls. Marianne went along to

The Breeding Period

34. Marianne became a Bunny Girl

be auditioned, and was one of the original eight girls to become a Bunny in London! Of course we were delighted; at the time it seemed a glamorous and prestigious job, certainly more financially rewarding than any job I had ever done, and on that score alone we were well satisfied. Unfortunately, Marianne's hours of work clashed with my own activities in music which I was loath to give up, so it meant paying out most of my earnings to baby-sitters. But we were still better off than before.

So for a while Marianne became the main breadwinner of our family, and I became the reluctant child-minder. Not that I didn't like spending time with Nicholas, but because it meant that I had to fulfil a role traditionally unexpected of the man of the family, I found it difficult to reconcile myself to this. Our flat was ideally situated between Holland Park and Kensington Gardens and so, although I had a choice of where to take our son for his daily exercise, I felt uncomfortable as the only male among the bevy of traditionally uniformed nannies and austere matrons who daily per-

The Breeding Period

35. *Nicholas at the Round Pond*

ambulated their little charges in the parks. I felt not only extremely incongruous but very frustrated, though there were compensations showing Nicholas the squirrels and feeding the gulls on the Round Pond or the Jays in Holland Park. In the evenings I played all the jobs that were offered me in the hope that sooner or later I would get an offer from a 'name' band. These jobs often involved long journeys to places as far away as Liverpool, Bradford or Newcastle, and although they paid better money it meant long hours of travelling, little sleep and poor food, and of course more to be paid out to baby-sitters. Luckily we had found a few ladies in our area who were reliable and trustworthy, so did not have to worry about whether Nicholas was getting proper attention.

Eventually I realised that the big-time offer was never going to come – the demand for jazz bands was slowly declining and the new sound of electric guitars provided by groups like the Shadows and the Beatles seemed to dominate the air-waves to an extent that effectively ruled out the chance of acoustic instruments making a come-back. I decided finally that I must forget the idea of earning a living from music and once again started to look for a daytime job. I still hoped to continue playing a few evenings a week in the London area, but did not wish to travel any distance.

As before, employers did not come rushing to avail themselves of my services, in fact I must have been even less likely to find a suitable job than ever. In the end I struck lucky: I saw an advertisement in one of the evening papers for an assistant gardener, no experience needed, in the roof-top garden of Derry and Toms department store in Kensington High Street, only five minutes walk from our flat. This astonishing place was previously quite unknown to me – an unlikely spot for an impecunious jazzman to spend his spare time unless in search of a job. But it was there that I found myself gainfully employed for the following year or so, and I would not have gained the gardening experience that brought me to Tresco if

The Breeding Period

I had not stumbled into Derry and Toms' Roof Garden on that auspicious day in 1962. I was agreeably surprised to find not only lawns and flowerbeds, but well-established trees, hedges, a stream and even a pair of resident Shelduck! The head gardener was a decent old chap who seemed quite impressed by my minimal knowledge of plants and gardening, enough at least to take me on, and could not have cared less about what I had been doing for the past few years. The only other gardener there was another ancient character called Paddy, with an Irish accent to match – he insisted, however, that he was a Londoner who had spent many years in Dublin as a younger man, having been at one time he told me one of the hated 'Black and Tans' at the time of the troubles in the 1920s.

My work in the roof-garden did not require much in the way of botanical knowledge as I soon found out. The first hour was devoted to sweeping the paths and tidying up, which in practice meant reading the paper and having a smoke for half an hour or so, then quickly tidying the paths before the 'Old Feller' as Paddy insisted on calling the head gardener turned up – just before the garden officially opened to the public. The day's work was then allocated; it usually consisted of cutting the various lawns, setting out bedding plants, lifting bulbs, carting rubbish and other basic maintenance tasks. The Old Feller did the tricky stuff like raising seedlings, striking cuttings and pruning while Paddy and I cleared up after him. This may sound pretty boring but it was a welcome

36. *Roof-top Shelducks*

change for me to be able to do some gainful work in the open air, and although the money was poor at least I felt that I was doing something to contribute to the family budget.

My musical career at this time was at a lower ebb than it had ever been, yet I still had an urge to perform if only I could find the right situation in the London area. One evening I went to the Marquee Club in Soho to hear a band that was beginning to create quite a stir on the jazz fringe – Alexis Korner's Blues Incorporated. Their music, earthy yet sophisticated, seemed to embrace the elements in music which most appealed to me, and I felt for the first time in years that I could contribute to what in Britain at least was quite a new musical direction. In the interval I got talking to a heavily bearded and impressive looking character who I soon learned was called Giorgio Gomelsky, a documentary film maker who specialised in jazz and seemed to know most of the people who mattered in the music business. Giorgio also had pretensions to becoming an impresario, and during the course of our conversation we decided to go into the blues business together – I would form and front a band, and he would be its manager. Rhythm and Blues we were convinced (and we were right) was the coming new trend, so we determined to be in the forefront of this musical revolution rather than chasing yesterday's trends, which I had spent the last few years doing to my cost.

Dave Hunt's Rhythm and Blues Band, as it was horrendously called, was destined to fail from the start – I attempted to fuse jazz, the only music of which I had any practical experience, with the rhythms of rock and boogie but it just would not work. I tried all kinds of musicians, from friends escaping from the rapidly crumbling jazz scene to ambitious youngsters with all the right ideas but not the beginnings of any musical ability. Some of those people are now household names in rock music; Charlie Watts of the Rolling Stones; Ray Davies, mastermind of the Kinks; and Lol Coxhill, eccentric jazz saxophonist, but in the context of my band they just got nowhere, and nor did I! Giorgio found a huge black singer called Frank who called himself Hamilton (Ray Charles) King. He was a good singer but had a hopeless stage temperament. Determined that we should succeed, Giorgio set about securing top bookings before the band had rehearsed enough material or a style of its own; he even set up weekly residencies for us in Soho and at Richmond before we had a steady personnel. Although we were only a part-time outfit I felt that the musicians should be reasonably rewarded for their efforts, but Giorgio insisted that the little money we earned should be ploughed back into publicity and advertising. Consequently most of what was available as wages went to the musicians leaving me with nothing! That of course was not the way to succeed as a bandleader, a profession which requires ruthlessness and greed as only two of the essential quali-

The Breeding Period

DIG THE **NEW** CRAZE IN BEAT MUSIC

RHYTHM & BLUES

AT THE

PICCADILLY JAZZ CLUB

41, GT. WINDMILL STREET, PICCADILLY CIRCUS

(Opposite Windmill Theatre)

EVERY FRIDAY

FRIDAY, 23rd   8 p.m. to 11 p.m.
DAVE HUNT'S R. & B. BAND
featuring HAMILTON KING
plus THE ROLLIN' STONES

FRIDAY, 30th   7.30 p.m. to 12.30 a.m.
**FIVE HOUR
RHYTHM & BLUES SPECIAL**
ALEXIS KORNER BLUES INCORPORATED
featuring U.S. SINGER, RON JONES
DAVE HUNT'S R. & B. BAND
with HAMILTON KING
plus THE ROLLIN' STONES
plus GUESTS

---

37. Top billing in the West End

ties, others being a sound business sense and skill in handling money.

An example of how a successful bandleader operates is illustrated by the following anecdote from my own experience. One day I received a phone call from a rival leader asking if I would do him a favour.

'My lads have all gone down with a mystery illness,' he explained, 'and we've got this residency at Blackheath – I don't want to disappoint the regular crowd,' he went on. 'Would you stand in for us please?'

'What's the job worth?' I replied, remembering to put business before sentiment.

The Breeding Period

'Well, you can keep all the door money less the hire of the hall. We usually get between three and four hundred people at three and six a head, so you should do all right,' he answered.

This sounded like a pretty good deal for us, so I agreed and after a few frantic phone calls had gathered some sort of band together for the evening. I should have smelled a rat – my rival's crowd prediction was cynically accurate. We pulled a 'crowd' of five people – more than three, but a lot less than five hundred!

Of course the other band had not been sick at all, but instead had been offered a more lucrative engagement. Next evening 'my lads' turned up at the Marquee, where the other band had a residency, and demanded a showdown – at least we got our expenses in the end.

Another group struggling to succeed at that time was a bunch of youngsters calling themselves the Rolling Stones. I was not very impressed with their music – it sounded to me like a bluesy extension of the skiffle which jazz musicians had always detested as an inferior brand of the real thing for people who could not play proper instruments. However, the Rolling Stones came cheap and were prepared to fill in our interval spots just for the chance to play in public, so I was quite happy to take advantage of their youthful enthusiasm. One day I got talking with Mick Jagger about the future prospects of our respective groups. My band had no hope of success he told me – after all I was almost thirty, and some of my band were even older he pointed out – how could we hope to score a hit with the younger generation at our advanced age? In view of his own youth and inexperience I refrained from correcting him, but I did begin to notice that the crowd around the bandstand at our Richmond gig seemed to swell during the interval, while our lads were at the bar and the Rolling Stones were performing.

To cut a long story short, in the desperately cold winter of 1962/3 I decided to give up the bandleading venture – I realised that I did not have the right temperament for it, and attendances during the snow were going from poor to pitiful, a state mirrored by my bank balance. So I gladly surrendered the Richmond residency to the Rolling Stones, who soon had to move to larger premises to accommodate their rapidly growing army of fans; severed my connections with Giorgio Gomelsky and retired hurt, though still not a lot wiser as far as the music business was concerned. In retrospect, the demise of the Dave Hunt Rhythm and Blues Band was a merciful release, but I did not see it as such at the time as I still desperately wanted to make a success of music somehow or other.

While all this had been going on Marianne was expecting our second son Martin, who was born just before Christmas 1962. This of course put a further strain on our resources, as Bunnies are not required to be pregnant, and Marianne had to give up the job – for a while at least. How we lived through that period I am not at all sure but it was certainly the lowest period of my life, which perhaps accounts for my not being able to recall too much detail. Suffice to say that we survived, I continued working at the roof-garden and as soon as she was able Marianne returned to her work as a Bunny, now at Raymond's Revue Bar in Soho. This was a less salubrious spot than the Bal Tabarin had been, and we were neither of us keen on it, but money had to be earned especially as there was now another mouth to feed.

Inevitably, after a few months away from music I got the urge to play again but this time I determined that it would be strictly for fun and that I would not get mixed up with management, heavy advertising or promotion. I got together with a few like-minded friends, all with secure jobs, and thus not financially dependent on music, and formed yet another group. Prophetically, perhaps, we hit on the name for the group; we called it 'The Ornithologists'. We played once weekly at a club in Earls Court. Billed as 'London's craziest bird-sanctuary' we played to a curious mixture of young Londoners, expatriates, students and debutantes, most of whom I guess lived in local bedsitter land. It was fun because we didn't take it too seriously.

However, this did not last for long either because the lease on our flat was due to expire shortly and we were faced with a dramatic rent increase if we wished to renew it. I decided that this was the signal for a drastic change, that I would somehow have to fight my way out of the London cage I found myself in and escape with Marianne and the children to a new life as far away from the pressures of life in the city as possible. But how I was to achieve this objective I had not the faintest idea.

One day I discussed my dilemma at work with Paddy. 'Young feller like you could get a gardening job anywhere in the world,' he said. 'Just get the *Gardener's Chronicle* when it comes out and look through the jobs vacant advertisements,' he advised. Not given to taking much notice of Paddy's advice in the normal course of events, I was still intrigued by the idea he had put into my head, so I duly purchased the next issue of *Gardener's Chronicle* when it appeared. Certainly there were plenty of jobs going, ranging from 'Single-handed gardener for country residence, wife to help with housework, live in' to 'Under-gardener of four required for commercial nursery'. Such jobs held no appeal, but there was one which did. It read something like this – 'Gardener required for luxury hotel on Tresco, Isles of Scilly. Rent-free cottage, wife to help in reception if desired'. This seemed too good to be true, though I'm sure I did not have much hope of getting the job. Nevertheless I applied and wrote a letter exaggerating my gardening abilities, enlarging on my horticultural experience, but playing down all my other activities of the previous few years. I then promptly forgot all about my application and carried on looking at further issues of *Gardener's Chronicle*, not with any serious hope in mind.

It came as quite a surprise, therefore, when I received a mystery phone call while at work. It turned out to be from the Island Hotel manager enquiring if I was still interested in the job, and if so was I prepared to come down for an interview at the weekend, all expenses paid? Well, a chance like that was too good to miss, even though I still didn't expect to be offered the job. I had never been to Scilly, but Father had once had a holiday there and I recalled having seen a film about the islands, so at least I knew that flowers and seabirds were among the island's attractions. I needed no further encouragement to escape from London for a few days, so gladly accepted the offer of a free trip to Scilly!

Before leaving London I made an effort to assume the appearance of what I imagined an aspiring gardener would look like. My hair was cut short and brushed into a parting which I would not have dreamed of having since leaving the army, and I had discarded my normal jeans in favour of a pair of baggy cords worn in conjunction with an old tweed sports jacket. Whether this was the right image or not I must have looked very peculiar, not at all like an erstwhile bandleader or former art student trendsetter! Whether my appearance helped I don't know, but the upshot was that I was eventually offered the job.

The hotel manager, Wing-Commander George Leatherbarrow, didn't seem to have much idea about gardening so I had little difficulty in pulling the wool over his eyes about my previous experience – in any case the garden did not seem too daunting a prospect and, although I saw it as a challenge, in my mood of optimism I felt quite confident I could handle it.

Deceiving the other Commander was a different matter though, and it is perhaps lucky that at that time I had no idea of the importance of Commander Tom Dorrien-Smith on Tresco. My first meeting with 'The Commander', as he was known to all the island residents, must have been contrived, though I was not aware of it at the time. On the Sunday it was suggested to me that while I was on Tresco I should take the opportunity of visiting the famous Abbey Gardens. I could think of nothing I would like better, so set off for an enjoyable morning. As I strolled around the garden in a state of near trance induced by the wealth of exotic flora, a stout gentleman with a rather grand manner wearing a blazer and a yachting cap engaged me in conversation about plants and gardens. During the course of the discussion I realised that he was clearly the owner, though at that time I was not aware that he was also involved with the hotel and effectively the ruler of the entire island. Anyway, he was obviously sufficiently impressed by our conversation to give the 'Wing-Commander' the appropriate signal for me to be offered the hotel gardener's job. I could hardly believe my luck – the only immediate problem as far as I could see would be uprooting my family from London and transplanting them on Tresco.

This proved to be no easy business but the hotel had offered to pay half our removal expenses, and a two-bedroomed cottage was to be prepared for our arrival, so the prospect of moving was not so daunting as it might have been. Consequently I was rather surprised that Marianne did not share my enthusiasm for the move. I could not understand why she was not overjoyed to be moving to what I envisaged as an idyllic existence! In the end I managed to convince her that we would all enjoy abundant sunshine and fresh air, that the children especially would benefit from individual attention at school and safety away from all the dangers of life in the city, so after a few weeks of hectic organisation we were ready to leave London and start our new life.

5. Post-breeding Dispersal

I arrived on Tresco a few days before the rest of the family, the idea being to get Number 1, Blockhouse Cottages in good order for their arrival. The interior of the cottage had been painted white throughout, covering the worst of the damp and blistered plasterwork, and although it looked bare without furniture at least it was habitable. Marianne had other opinions though. 'Don't bother to unpack, because I'm going back with the children on the next boat,' she said after one look at the place. Luckily there was no boat for a couple of days and during that time the sun shone, the boys discovered the beach and Marianne resigned herself to the primitive conditions, having been promised certain improvements though some of these took a while to materialise.

I became quickly involved in my new job and soon discovered that it was going to be a lot more demanding than I had at first thought. I began to realise then why I had been offered the job in the first place – because no experienced gardener in his right mind would have taken it on! The remote situation, low wages and dubious perks like a tied cottage and free milk would have put off anyone who had taken time to think, but I was neither an experienced gardener nor a very good judge of situations, though I resolved to make the best of it – what else could I do?

The job was very different from the one I had left behind at Derry and Toms. For a start I had nobody to give me daily instructions apart from the Wing-Commander, who seemed quite content to let me get on with it. So long as the drive was kept clean and tidy, the lawns mowed, the hedges trimmed, the weeds removed and the greenhouse watered my time was my own once I had made sure the deckchairs were repaired, the croquet set and the bowls put away at the end of the day. True I had a pensioner to help me for a few hours each week, but I spent so much time tidying up after him that I might as well have done his work myself. Needless to say I was kept busy, and had little time or energy left for the family in the evenings. Even on Sundays I found I was expected to be on call after church, when the Commander made his weekly inspection of the hotel and grounds.

However, after the initial shock of actually having to work for a living, I eventually learned how to get the job done within a normal working day, and by the autumn was even able to find some time to go bird-watching.

I soon found out that on Tresco, the Abbey Pool and nearby Great Pool were the best habitats for my favourite birds – waders and waterfowl, and I revelled in the chance to wrestle once again with the identification problems that these groups presented. I was of course extremely rusty in this department, but it was not long before I had refamiliarised myself with the commoner species and was hot on the trail of something different. It was as if the intervening years had never happened and I felt quite rejuvenated. I was also a lot fitter physically than I had been since coming out of the army, for the years of night-clubs, lack of exercise and beer-swilling had inevitably had their effect. Now I could not afford more than the occasional half-pint and was getting all the fresh air and exercise I needed. No wonder I felt different!

The children also seemed happy in their new environment, and were unquestionably fitter and healthier than they had ever been in London. Money was in short supply though – Marianne earned a little extra each week by arranging flowers at the hotel, but our joint earnings could not compare to what we had been used to in London even when times had been hard! Still, our outgoings were small and somehow we managed to get by.

Our finances were certainly not up to buying me a new pair of binoculars; my ancient ex-army glasses were the only optical aids I possessed. I remembered that Father had an old three-draw telescope he rarely used so I wrote and asked him for it to help me scan the beaches and distant rocks beyond the range of my binoculars. The day the telescope arrived I could hardly wait to put it to some use, so as soon as I finished work I hopped on my bike and made for the beach at Carn Near which usually had the best selection of waders. The beach was pretty deserted, so I trained the scope on what I assumed was a gull on an off-shore rock. As I focused I had the surprise of my life because not a gull but a Snowy Owl materialised! It was as if I had received a magical device rather than an optical aid, but there was no doubting the reality of the Snowy Owl, which I watched as it flew from the rocks to the shore and eventually disappeared in the direction of St Martin's. That evening I phoned the St Agnes Bird Observatory where I knew some ornithologists were staying who would surely be excited at my news! 'Oh it's still about is it?' came the deflating reply at the other end of the phone. Apparently it had been seen on St Agnes

38. *'A Snowy Owl materialised'*

the previous day, so I had not made a new discovery. Despite this, nothing could rob me of the elation I felt after seeing my first genuine Scilly rarity, even if I could not take the credit for it.

Next day I had the chance of spending the afternoon on the island of Bryher and jumped at the opportunity to widen my horizons. I found Bryher less 'sophisticated' than Tresco, and I enjoyed especially the rugged aspect of the view looking westwards across the Norrard Rocks. In one of the overgrown fields close to the pool I spotted a bird that was totally unfamiliar to me; at first I could not even assign it to any obvious family. However, it was keeping company with a group of starlings, and I soon noticed that its structure and behaviour were similar, though unlike the starlings it was a pale sandy colour with darker wings and tail. I concluded it was some kind of abnormal starling and thought no more of it. Of course nowadays such a bird would be immediately identified as a juvenile Rose-coloured Starling, a species which has been recorded almost annually on Scilly during the last few years. For me then it was a puzzle until a few days later when I was invited by the Commander to view his family collection of stuffed birds which in those days was still housed in Tresco Abbey. Among the many specimens in glass cases which lined the corridor was a bird identical to my Bryher mystery, alongside a fully adult specimen of the same species, resplendent in pink and glossy black plumage. So I no longer wondered what on earth I had seen a few days earlier.

There were also many other more impressive birds in the collection, among them two handsome Snowy Owls, Gyr Falcons, an American Nighthawk and many other rarities too many to detail here – and all collected by the Dorrien-Smith family during their tenure of Tresco, as the Commander took great pride in explaining.

Which led, I was soon to realise, to the main reason why I was in the privileged position of being shown the collection in the first place – 'As it seems you're a keen and knowledgeable chap about birds,' said the Commander, 'I hope you'll let me know straight away if you find any rarities. Then,' he continued, 'I can send out Fred Wardle who'll shoot it for the collection.'

Post-breeding Dispersal

I was appalled – though relieved to think that the Snowy Owl seemed to have disappeared. I certainly would not reveal its whereabouts if I saw it again and resolved to keep my mouth shut about any other rarities that might turn up in future, at least until I felt certain they had moved on.

Fred Wardle was the estate gamekeeper, and at that time an unknown quantity as far as I was concerned. I later came to know Fred very well and found that although he had shot a number of the specimens in the collection, including a Black Kite in 1938 and a Honey Buzzard as recently as 1958, he no longer had any heart for it, preferring to see birds of prey flying wild and free. Fred, although he had lived on Tresco since he was quite a young man, still retained a North Country accent which revealed his upbringing and early experience in the woods and fells of Cumbria. He was a delightful character with a lifetime's experience of his job and therefore considerable knowledge of wildlife – he proved to be an invaluable contact and often informed me of any unusual birds seen in the course of his duties. His knowledge of ducks and raptors was considerable though he was a little shaky on the subject of waders and small birds, but this he freely admitted and we often had long chats about bird identification problems. Sadly Fred passed away not long after his retirement in the early seventies, leaving a gap on Tresco that has never been satisfactorily filled.

The Snowy Owl had not vanished as I had half hoped, but eventually stayed around Tresco and the neighbouring islands for the rest of the winter though it had nothing to fear from Fred as I have just explained. For a while Snowy, as we took to calling her, made a habit of roosting in a tree within sight of my place of work and was an almost daily acquaintance. Another wanderer from the Arctic which appeared during that memorable first winter on Tresco was a magnificent Gyr Falcon of the Greenland race. I had the thrill of watching that spectacular bird as it hung in the wind over Old Grimsby harbour on Christmas Day, viewing through our kitchen window!

39. *A Gyr Falcon for Christmas*

That winter I also first met several bird-watchers who were to become regular visitors to Tresco in future years, among them Bernard King who has remained a Scilly enthusiast ever since and with whom I have shared the enjoyment of many birds, rare and otherwise, among the islands.

Having survived our first winter on Tresco we looked forward to the summer of 1965 and whatever it would bring, but especially sunshine and a chance to explore as the previous summer had been spent just accustoming ourselves to our new way of life. I had got hold of a small rowing boat in which I planned to visit some of the neighbouring islands, and hoped that the family would enjoy picnics and a chance to explore a bit as well.

One morning in the spring of 1965 I woke to hear the unmistakable trilling of a Grasshopper Warbler, a sound I had not heard for many years. I rose to find a still, but misty morning, the undergrowth dripping with moisture, and the reeling of more Grasshopper Warblers coming from every direction – it seemed as if every bramble brake held a singing bird. Sedge Warblers could also be detected as my ear tuned in to the different songs – it was my first experience of a real 'fall' of spring bird migrants. Later that day I was surprised to hear a Nightingale in full song as I returned from work at lunch time, and it was still singing in the same place that evening. After I finished work I went for a walk along the footpath leading from the Old Blockhouse past Borough Farm towards the pools, and on my way flushed at least six Hoopoes from the sandy fields or along the path. These were by no means confined to Tresco, however, and the *Cornwall Bird Report* later published an estimate of no fewer than seventeen Hoopoes among the islands for that spring – a total that has never been surpassed in my experience.

Inevitably not all my observations took place outside working hours, and I used to keep my trusty but battered old binoculars

40. *Unprecedented arrival of Hoopoes*

round my neck or close at hand wherever I was at work. This naturally caused much comment among the visitors and guests at the hotel and I got used to comments like 'Help to find the weeds do they?' or 'Seen any bare birds lately?' My work suffered from many diversions that summer and I seem to remember spending more and more time away from the lawns and hedges, while attending to the greenhouse where I found many extra jobs seemed to need doing! Whether the Wing-Commander was aware of my shirking or not I'm not sure, but luckily for me the state of the hotel garden was only one of the manager's minor problems.

Meanwhile the guests seemed amused to find that the hotel gardener was also knowledgeable about the local bird life. I soon found those among them who would like to chat about birds, flowers and whatever else interested them – hindering the work in progress at the same time. Not that I worried a lot – as far as I was concerned if I was entertaining the guests it was all part of my job, and the task in hand could always be completed tomorrow! I was often asked for a few cuttings or seedlings of plants which took the guests' fancy, and usually able to oblige. Although I never set out to sell these I was likely to be rewarded by means of a tip, sometimes in excess of the value of the material, and I soon learnt to accept these extras as part of the job as well!

In fact, to have been able to live off my wages alone (I was earning a basic £10 per week) would have been impossible. Marianne started to make up shell necklaces and other crafty items for John Hamilton, a former manager of the hotel who had set up a workshop, purveying what were advertised as 'Gifts from the Sea', most of which were assembled by island wives – the work was time-consuming and not very remunerative, but it brought in a few extra pounds, and with two growing boys to feed every little helped.

I also started to put my art-school training to some use after those wasted years, producing small water-colour bird paintings which I was allowed to display in the hotel lobby. To my surprise I found a few of the guests were prepared to buy them at two or three guineas a time – another useful supplement to my pitiful pay packet. Encouraged by this I turned my hand to larger pictures of local landscapes, for which I could get as much as a fiver – suddenly we seemed almost well off!

Nicholas had settled in quite well at the local junior school and made a few friends among the other island children after the initial strangeness, but Martin was proving a full-time headache for Marianne, being at the age when he was continually in need of attention. Although Tresco was safe for young children in the generally accepted senses concerning traffic and molestation there were other hazards, especially the threat of the sea. When on one occasion Martin vanished inexplicably, while Marianne was doing her daily shopping, we spent some anxious moments scouring the

Post-breeding Dispersal

41. Early morning at Men-a-Vaur

beaches until he was eventually discovered having a wonderful time dismantling the Raeburn stove in the cottage next to our own which was untenanted at the time. But apart from such occasional worries our life on Tresco, if not quite as idyllic as I had fondly imagined it would be, was nevertheless acceptable and made all the more pleasant by an exceptionally good summer.

On fine Sunday mornings I would rise early before Marianne or the children were awake, and row out in the boat to St Helen's or Men-a-Vaur, where I found gulls, terns and even Puffins nesting. I had hoped to encourage the family to join me for boating trips to the other islands but the size of my skiff did not induce much confidence in Marianne, and I must admit that it was a tight squeeze trying to fit us all in at the same time, so those plans were shelved for that summer at least. Living as we did almost on the beach the children did not need to look far for sand or sea and could have as much fun there as they wanted and still be within earshot of our cottage.

In August 1965 I discovered my first American bird rarity on the Tresco Great Pool. I had slipped away from the annual Church fête for a few minutes to have a quick look at the waders on the shore, among which was one clearly smaller than even a couple of Little Stints nearby. I made as detailed a description of it as I could in the circumstances, which were not too good as the light was not favourable, and after consulting the limited literature available to me decided it must be what in those days was known as an American Stint (now Least Sandpiper). Next day I saw it again and satisfied myself that my identification was correct, and my observation was finally published the following year in the *Cornwall Bird Report*. Details were also forwarded to the Rarities Committee of *British Birds* magazine, but in their wisdom the BBRC as I will refer to that august body of gentlemen hereafter (some call them the Ten Rare Men, others are much less complimentary!) found it unacceptable. Their original assessment was endorsed by a more recent committee, and although I am personally quite happy with the record am prepared to accept their decision. However, readers may care to judge for themselves, and I include the description as submitted to BBRC for interest, and NOT I hasten to add because I am hoping for a reversal of that decision which I know was made in good faith and with the objectivity required by a committee vetting records of national importance.

Tresco. On 24th August 1965 a bird was first seen in company with two Dunlin and two Little Stints feeding at the S.E. end of the Great Pool. It was observed in good evening sunlight between 7.15 and 7.45 p.m. at very close range, sometimes as close as 6-8 ft. It was very tame and although repeatedly flushed always returned to the same piece of shore.

In size it was noticeably smaller than the stints, though the difference in size was not as marked as between stint and Dunlin. Its general colouring was a greyish brown, as opposed to the more rufous stints, and nearer to that of the Dunlin. There was no trace of the characteristic pale 'V' marking on the mantle which was pronounced in both stints, and this uniformity was carried up to the head and neck which were similarly streaked greyish brown, as also were the throat, upper breast and flanks. Belly and under-tail coverts were whitish. The bill was black, short and typically stint-like. The legs appeared a dark olive green. When at very short range I scrutinised the bird's feet for any sign of partial webbing between the toes, but none was apparent.

When flushed the bird uttered a sharp 'Jeez' sometimes twice repeated, rather reminiscent of a feeble Dunlin. In flight it was not easy to discern much detail as the bird invariably flew off towards the sun, which at the time was low in the West; but it appeared similar to the stints, i.e. a faint wing-bar and white outer tail-coverts.

The next day I saw the bird again in strong sunlight on the shore of the nearby Abbey Pool. I was able to confirm all my previous day's observations, though in the stronger light the legs did not appear as dark as in the evening, and would be better described as just olive green.

It should be noted that the preceding few days had been very unsettled with N.W. winds often reaching gale force.

And that is what it took to convince the BBRC that I had NOT seen a Least Sandpiper! Actually I have abridged the original somewhat to avoid boredom! But as this helps to illustrate how I began to fall out with the BBRC after my first report, I thought it worth including.

A few days later I discovered a White-rumped Sandpiper in the same place, but as this bird was later trapped and ringed by two reputable ornithologists that record could not be refuted. Julian Rolls and Richard Frost Lee were among the first serious bird-watchers to visit Tresco during my time there, and they made several return stays in the years to come. Their trapping efforts on the shore and in the reed-beds and farm land of Tresco led to some useful information being added to the meagre knowledge we had then of both common birds and unusual ones visiting Scilly.

So that was how the ornithological establishment came to be aware that a new observer had appeared on Tresco which until then had been rather neglected, as most visiting bird-watchers headed automatically for St Agnes and the observatory there. Not that my sightings were so remarkable – merely highlighting what should have been obvious, that all the islands had potential for producing records of unusual birds, especially Tresco with its rich variety of habitats. Nevertheless, for a while it seems my observations on Tresco were regarded as dubious to say the least, and unless corroborated rarely found their way into the record books.

Meanwhile the majority of rarities accepted by the BBRC from Scilly were ones reported from St Agnes, so at the end of October 1965 I decided to spend a weekend at the Observatory to find out what it was about St Agnes that attracted not only the majority of bird-watchers, but the birds as well! Getting to St Agnes from Tresco was not so easy in those days, involving first a trip to St Mary's in order to catch the daily supply launch, the timing of which depended on variable factors like tides, and the arrival or departure of the steamer from Penzance. Having wasted half a day in that process I arrived on St Agnes some time after midday to find most of the observers packing up preparatory to departure on the afternoon steamer. However, a few were staying on including Brian Milne and Andy (or was it Tony?) Sudbury who were to be my companions for the remainder of the weekend.

The leaving bird-watchers included none other than D.D. Harber, who by then had risen to such eminence in the ornithological world that he was secretary of the BBRC and a force to be reckoned with! In fact I suspect he was a major factor behind the failure of several of my records to be accepted, and he made no secret of his lack of faith in me and my equipment.

42. *St Agnes' famous lighthouse*

'Look at his telescope – the object lens is the diameter of a penny! How can he expect to see anything through that? And his binoculars – wartime rejects!' were his comments to Ron Johns, another regular on St Agnes in the sixties.

To be fair, Harber and the others did help me to see the Spotted Sandpiper which for weeks had been frequenting the beach at Priglis, the sheltered bay on St Agnes' western shore, though since it was calling continually and drawing attention to itself in no uncertain way, I would have had to be deaf or very ignorant not to have found it unaided! Cliff Waller, one of the original discoverers of this bird in September, relates a series of events involving the identification of the Spotted Sandpiper worthy of being repeated.

The initial identification stemmed from the bird having been trapped by Cliff and others, following suspicion that it might not have been a Common Sandpiper because of its unusual call. The identification hinged on various fine points of plumage detail too complicated to go into here, but after examination the observers were satisfied that it was indeed a Spotted Sandpiper and released it, having made the necessary written description. Harber, who arrived on St Agnes a few days later, was not satisfied with their original diagnosis and insisted upon the unfortunate bird being re-trapped and examined yet again. It was photographed and even some of its feathers were removed for comparison with museum skins before it was finally released. As the observatory log laconically reported that day ... 'the remains were flung on the beach,

43. *'It's a yankee cuckoo!'*

and immediately resumed feeding happily'. When I saw it a couple of weeks later its identification pointers had of course been removed, but it was exceptionally vocal that evening and the following day was gone – none the worse it seemed for the less than considerate way it had been treated in the furtherance of establishing a record! I am glad that the techniques of field identification have improved so much that today such measures would not even be considered.

As Harber and most of the other observers left later that afternoon, they gleefully reminded us that it was now late October and that we could not expect much more in the way of migrants as the season was now over. And they seemed to be right for St Agnes was apparently deserted by all but the obviously resident species next morning.

The few of us remaining combed the island unsuccessfully for new arrivals, ending up on The Gugh where Harber had discovered a Rustic Bunting some days earlier. As we stood by an overgrown hedge in what remained of the area that had once been a flourishing farm, discussing the absence of birds in desultory fashion, I noticed a bird's head, with decurved beak and yellow-rimmed eye, peering warily out from the hedge only a few feet away. I hardly had time to draw attention to this before the bird burst from cover and flew down the slope, revealing a slim grey body, long tail and white underparts. There was also white in the tail, and the wings showed some rufous, but this was all we could see before the bird buried itself in another hedge. 'It's a yankee cuckoo!' shouted one of my companions, and from what we had seen this seemed a likely conclusion, though we would need to see more of it for a firm identification, so while someone followed the bird to the region in which it had vanished, the rest of us watched for it to reappear. Sure enough it soon obliged and on this occasion flew across the bay and settled, clinging to the vertical face of a rock. There we were able to confirm our previous impressions of the Yellow-billed Cuckoo which it proved to be, before it once more flew off, this time heading for St Mary's at great speed and gaining height continually until it disappeared from sight. This was the first record of the species in Scilly since 1940, though there have been several since. But so much for the migration season being over, and I returned to Tresco on the Monday with renewed enthusiasm, determined not only to keep scouring Tresco but to return to St Agnes as often as possible in future, though I knew this could not happen again until the following year.

A few days later my dream of finding a great rarity on Tresco was realised at last, but on that occasion there was no one with whom to share my excitement and I was obliged to keep the news to myself for fear of the bird being bagged for the Dorrien-Smith collection. I did phone Brian Milne, who was still staying on at St Agnes, but there was an easterly gale raging and no chance of a boat between the two islands.

I found the bird in question during the course of a routine visit to the Abbey Gardens to scrounge some cuttings for the hotel garden. I knew all the gardeners quite well by then, but could not take them into my confidence due to my fears for the bird's safety – and it was on extremely dangerous ground! It frequented the area of the Abbey Gardens' rubbish dump, where I first discovered it while taking a short cut home through the woods. Obviously a thrush, but with large crescentic markings both above and below, I knew at once what it was for I had often looked at the picture of White's Thrush in the Handbook, and fantasised about one day seeing it. But this was no fantasy, and I was able to watch it feeding almost at my feet, as it forked over the decaying material on the dump. Since there was no one I could trust to corroborate the record, I determined to make as detailed a description of the bird as possible. Knowing that the species should show a very distinctive underwing pattern, I attempted to check this feature but without success for whenever I flushed the bird it flew off low into the surrounding woodland without giving me a chance to view its underwing at all. Obviously I did not wish to keep on disturbing the White's Thrush, so in the end had to content myself with a detailed account of what I did see, and an honest admission that the underwing pattern had not been observed. As you have probably guessed, the record was found unacceptable by the BBRC 'because the underwing was not

44. White's Thrush – the fantasy realised

seen'. I supposed that Harber either presumed that I could not tell a young Mistle Thrush when I saw one, or thought I was making it up! I was becoming rapidly disillusioned by the attitude of the committee. To get the record put straight I recently resubmitted my sighting to the current BBRC – they accepted it at the same time as they turned down my Least Sandpiper, so I cannot claim this as personal prejudice as I am afraid I did in the case of Harber's committee.

Another unusual visitor in November that year was a Rough-legged Buzzard which flew over Old Grimsby one day and headed for St Mary's. This was later confirmed by Cliff Waller, who had returned to Scilly for the winter, having found a job as a butcher's assistant on St Mary's. Sadly for me, because I could really have done with a few kindred spirits resident on other islands, Cliff did not stay for long and I was soon on my own again.

Not that there were no other bird-watchers living on Scilly at that time, but those that were seemed rather remote or aloof figures of an older generation and, in any case, I had little or no contact with them until later. Miss Hilda Quick lived on St Agnes and, although a genuinely delightful person as I was to discover in time, was an old lady who spent most of her time pottering about in her garden. Her eyesight was failing and apart from the large and conspicuous rarities which appeared virtually on her doorstep, she was not likely to be much of an ally in corroborating my sightings. The others lived on St Mary's – Peter (P.Z.) Mackenzie was the islands' vet and a small-time flower farmer who seemed to look rather askance at the enthusiasms of younger bird-watchers, and certainly did not encourage communication at first with his rather offhand manner. Ron Symons, a local man who worked as a docker on St Mary's quay, and still does, was not known to me at all until I finally came to live on St Mary's, and that was it!

So the following year I was still a lone observer on Tresco and when I spotted a large raptor approaching as I was working in the kitchen garden which I had no doubt was a Black Kite, I had little hope of the record being accepted by the BBRC, indeed I had almost decided not to submit it until corroboration came from a quite unexpected quarter.

Derek Upton, plant propagator in the Abbey Gardens, told me that he had seen a bird which on referring to his field-guide he believed to have been a Black Kite. On comparing notes of time and place it became clear that he had seen the bird only a matter of minutes before I had, and his description tallied in every detail. Although not a keen bird-watcher Derek had been a member of the Young Ornithologists' Club while at school, and his confirmation of my identification was just what I needed to get the record accepted! Except that after deliberation, the BBRC rejected it! That was the last straw and I resolved not to submit any future records. As far as I could see Harber and his committee would never believe any of my Tresco observations unless confirmed by bird-watchers known to them, so I felt I was wasting time and energy in trying to prove anything to them.

It was John Parslow, at that time the secretary of the St Agnes Observatory, who eventually persuaded me to change my mind about this – I re-submitted our Black Kite record, adding that I had had previous experience of the species while living in Germany and it was finally accepted, becoming the sixth British record and

45. 'That bird belongs on Bryher'

Scilly's third – the first having been shot by Fred Wardle back in 1938 and another seen in 1942.

That spring also produced a few other exciting birds, including a Purple Heron, two Garganeys and a Little Egret, both the latter species having astonishingly spent much of their time frequenting an area of flooded grassland close to Tresco church. Over the coming years I found that this tiny area, after the winter rains, invariably attracted birds, sometimes maybe a Greenshank, a couple of Ruffs or even a flock of Black-tailed Godwits on one occasion.

Returning to the Little Egret, which obligingly came regularly to that little bit of flood water. I had recently got hold of a simple reflex camera and, anxious to put my new acquisition to the test, attempted to sneak up close to the egret (I had no such refinement as a telephoto lens), making use of a low wall and a fold in the ground for concealment. Just as I was within range of my subject, I was horrified to see a man approaching the egret, his arms flailing, crying 'shoo, shoo'! Not surprisingly the bird rose in the air and flapped away over the hill behind the church. I too rose from the grass where I had been concealed. 'What did you do that for?' I cried, furious. 'That bird belongs on Bryher,' replied the recipient of my wrath. The remainder of our conversation is not printable, but for some reason the lunatic who chased off the egret, and was staying on Bryher, where it shared its feeding time, had decided it was not intended to grace any other island and had taken appropriate action. I never saw the egret again on Tresco but whether it returned to Bryher as it had been exhorted I was too disgusted to find out!

In May of my third year on Tresco, a large arrival of House Martins took place. I was interested to notice that they were attempting to nest on several buildings not normally favoured by the species, including the Island Hotel. Anxious that they should succeed, a considerable amount of my work effort was devoted to creating and maintaining muddy puddles along the hotel drive so that the martins would be able to find sufficient nesting material during the rather dry weather that prevailed at the time. This caused some amusement among the guests, but not I guess the management, and it was about this time that relations between me and George Leatherbarrow started to come under some strain.

To compound this situation, I began to realise that my attempts to improve or add to the existing garden by trying out new plants or creating miniature cactus gardens were not really appreciated. The Commander had already explained to me in no uncertain terms that it was not my job to attempt to vie with the Abbey Gardens, and it came home to me that really all that was wanted was a maintenance man. My enthusiasm naturally waned at this point for gardening of this kind did not stimulate or interest me, so I found little to occupy my mind except to devise ways of making the job less boring – and of course there were always plenty of diversions!

I had agreed, rashly in some respects, to undertake the production of selected vegetables for the hotel kitchen but this at least gave me an excuse to escape from the managerial eye for hours at a time, since the kitchen garden was separate from the hotel grounds and hidden behind a series of hedges; and here I raised lettuce, radishes and other easy crops as a relief from cutting lawns, trimming hedges and other chores which I found more and more oppressive. The only problem was, the more I neglected the hotel garden, the more guilty I began to feel and the more I needed to hide away from the managerial wrath by immersing myself in tasks that were not really necessary in the kitchen garden!

While I was thus lurking out of sight and trying to think of good excuses not to be mowing or trimming, I became aware that a pair of Lesser Whitethroats was nesting in one of the fields close by the kitchen garden. Naturally, for this was a first record of the species breeding anywhere in Scilly, I did not need encouragement to waste even more of my employer's time in watching the birds' progress.

During my spare time my boating endeavours continued as I had now purchased a larger and therefore safer rowing boat, or 'punt' as they are always termed in Scilly. On calm days I even risked rowing out to Men-a-Vaur, the isolated rock stack to the north-east of Tresco where Guillemots, Razorbills and Fulmars were among the nesting birds. It was a wonderful feeling to be there in the early morning when the sun was just warming up the rocks and the birds were going about their daily routine apparently quite unconcerned by the presence of me and my boat gently nudging close to

46. *Singing Lesser Whitethroat*

Post-breeding Dispersal

their breeding ledges. I could easily forget any problems that might be worrying me at work or at home in such surroundings, in the company of the seabirds, and rarely missed the opportunity to go there if it was calm enough.

Being still supremely ignorant of the local hazards it was a shock one morning to see the rocks of the treacherous reef known as the Golden Ball gliding only a few inches below the bottom of the boat! I had not realised I was over it in my blissful ignorance. On another occasion following a family picnic on St Martin's, where I had finally persuaded Marianne and the boys to accompany me, I had extreme difficulty in rowing back against the tide which of course had been a big advantage on the way across, but I had made no allowance for that!

Despite these setbacks I refused to allow myself the luxury of an outboard engine, preferring to sweat in silence. How we survived that summer without the lifeboat having to be launched to save us I don't know, but we did, though the family did not seem to be as enthusiastic about boating as I was by the end of it!

Seabirds and boats lost their appeal for me as well when the autumn came round again. It was not long before I had added a Baird's Sandpiper to the list of American waders I had found on Tresco, and this was soon followed by another White-rumped and Tresco's first Spotted Sandpipers. All of these were seen well and 'vetted' by ornithologists of repute including John Parslow, Ian (D.I.M.) Wallace and Peter Mackenzie, with whom I was at last beginning to feel comfortable, having learnt that his rather off-putting manner was not intentional. Peter was later to become a good friend in many ways, but at this stage he was just a useful contact on St Mary's who I found was now prepared to help me in substantiating my sightings and thus well worth cultivating. I had also become more acceptable in the eyes of the ornithological establishment following these discoveries and having met and talked with Ian Wallace, at that time a well-respected member of the BBRC.

It still irked me, though, that certain of the small migrants of European origin which were of regular occurrence on St Agnes never seemed to appear on Tresco. I could not believe that Bluethroats, Wrynecks and Red-breasted Flycatchers bypassed Tresco completely, but despite continual searching those species continued to elude me. If I had really thought about it I would have realised just why I was not finding them, but at the time I was still very ignorant about the different habitat requirements of different species and very much rooted in a certain type of bird-watching routine which did not allow for much deviation.

As soon as I finished work I would leap on my bike and pedal madly to the pools while there was still enough light for observation. By the time I had surveyed the pools and beaches it would be getting too dark to continue, so my bird-watching for the day

Post-breeding Dispersal

47. Bluethroat from the bathtub

was over. This was fine for finding waders, but it meant that the prime habitats for small birds were being largely neglected, apart from the hotel garden, which got a good surveying first thing each morning, but was a far less attractive habitat for small bird migrants than the sallows surrounding the Great Pool or Tresco's extensive areas of woodland, the Abbey Gardens and farmland which I seldom had the chance to visit. So it was small wonder that one person (myself), working Tresco in my spare time, could not find the birds which St Agnes (an island with much more restricted cover) produced when scoured all day by a handful at least of keen and dedicated observers with nothing else to think about.

One night in late September I decided to take a bath on my return from an evening's bird-watching. After first hearing someone knocking on the front door, followed by Marianne's voice in conversation with the caller, there was a tap on the window of the bathroom, which was actually at ground level in one of the adjoining outbuildings. Puzzled, I opened the window to find a hand thrust towards me holding a fine male White-spotted Bluethroat! My friends Julian and Richard who had been mist-netting at dusk in the reeds close to the Great Pool, had trapped the Bluethroat and remembering my 'need' to see one had kindly obliged by bringing the bird to me, thus breaking the first jinx in that trio of birds which had escaped me for so long. I had to wait until the following year for the others, but the spell had been broken, and I doubt if any other bird-watcher can claim to have seen their first of any species while sitting in the bath tub! The Bluethroat was overnighted in a ringing bag and returned to the reed-bed at dawn. I was there to see it go because I could not count it until I had seen it flying free. It disappeared, none the worse for its experience, by burying itself in the reeds once more and I realised that this was why I had not seen a Bluethroat before on Tresco. Similarly a rare warbler or flycatcher only needed to dive into the woods to become equally invisible – no wonder I had so far failed in that direction as well.

Before that autumn was out, though, I had the rare chance of being in on a bit of ornithological history-making. But it would not have been possible without the presence of my old friend Bernard King. Bernard and his wife Marjorie were staying at Tresco's New Inn for a few days in October, and on the Sunday morning I called on them at breakfast time to say 'hello'. Bernard was in quite a state!

'My dear chap, I'm so glad you came,' said Bernard, 'because I've seen a warbler by the pool, and I've no idea what it is.'

Coming from Bernard, who I knew as a bird-watcher with a great deal of experience, this was a very unusual state of affairs, but as he went on to describe the bird I realised why he was puzzled. I wasted no time in getting to the spot where the mystery warbler had last been seen and, after some frantic searching of the sallows along the track known as pool road, finally got a series of glimpses of what was clearly the same bird – I was not surprised at Bernard's failure to identify it. Although it was hyperactive, only giving the briefest of views, it was possible bit by bit to piece together a detailed description as follows:

Size about that of a Chiffchaff, though in shape more like a large *Regulus*. General colouring of the upperparts bluish grey, but on the mantle a distinct tinge of yellowish green with a hint of similar suffusion on the crown. The forehead appeared a slightly paler grey. Eye dark, with a crescentic white mark above and below it suggesting feathered 'eyelids'. Wings blue-grey with two short but very prominent white wing-bars formed by broad white tips to median and greater coverts. Rump and tail blue-grey with outer tail feathers appearing a shade paler. Throat and breast vivid yellow, more intense on the sides of the upper breast. Rest of underparts pure white. Legs a bright pinkish-flesh colour and bill similar though darker in tone.

It was clearly not of European origin, being by its very nature much brighter and more eye-catching than any warbler I had ever seen before, so the conclusion followed that it must have somehow crossed the Atlantic from North America, though in those days

Post-breeding Dispersal

48. *Britain's first Parula Warbler*

there were very few if any records of such a tiny bird having survived that crossing unaided. However, since Bernard had Peterson's *Field Guide to Western Birds* with him, we pored through it exhaustively. Only one possibility emerged – something called an Olive-backed Warbler, but since this was only supposed to occur in the Rio Grande Valley of Texas, or in Central America, this seemed a very unlikely candidate. Looking at the book again I found, in the small print of the Appendix, reference to the very similar Parula Warbler with a distribution that included Ontario and Nova Scotia, and thus far more likely to have been blown off course on migration. It was of course the first occurrence of the species in Britain, though it had been previously recorded in Iceland so was not new to Europe.

That evening a phone call to St Agnes brought a hardy handful of eager bird-watchers to Tresco on the following day, despite a quite severe westerly gale. Among them were Roger and Liz Charlwood, Paul Holness and Ron Johns, most of them still regular visitors to Scilly at migration time. They all managed to see the Parula Warbler and concurred happily with our identification. I remember hearing the refrain 'Have you seen a Parula Warbler lurking in the shadows?' being very similar to a Rolling Stones song of the same period. It certainly caught on with the St Agnes Observatory crowd that October! I believe also that this was the first occasion on which a boat was hired for the special purpose of bringing bird-watchers from one island to another for a single bird, creating a

Post-breeding Dispersal

49. St Agnes Observatory

precedent to what has since become a commonplace event.

The remainder of that autumn was rather an anticlimax; even my eagerly awaited weekend on St Agnes failed to produce any unusual birds, though I got to know more of the visiting observers and some of the local inhabitants as well. The Observatory was housed in the farm buildings at the foot of the hill below the lighthouse. There, in dormitory accommodation for no more than about eight people, with basic cooking and sanitary arrangements, one could stay for four shillings and sixpence per night, which included milk fresh from the cow! Lewis Hicks, the farmer who kindly provided these facilities still lives in the lighthouse, but his son Francis, himself a keen bird-watcher runs the farm. For a variety of reasons, one of them the need to show a profit, the old farm buildings are no longer an observatory but a self-contained letting cottage much in demand during the holiday season as well as at migration periods. Bird-watchers still stay regularly on St Agnes in autumn but can no longer be accommodated all under one roof, being scattered about the island in various kinds of lodgings from guest houses to the camp-site. Now that other islands have been found to be as good as or sometimes even better for birds, the emphasis has shifted to St Mary's as the main place to stay. But the atmosphere of those early days and the enthusiasm generated by the visitors to the St Agnes Observatory will linger in my memory for ever, and I suspect in the memories of all the others who recall those early days – when a group of more than a dozen bird-watchers constituted a crowd, and a rarity could be enjoyed by all without disturbing either the bird, the local residents or one's fellow enthusiasts! Sadly the Observatory had to close and since the late 1960s no organised ringing has been done on St Agnes except by the occasional individual or visiting group.

Another winter came and went, but in the spring of 1967 a dramatic event took place which might have had a truly disastrous effect on Scilly's seabird population – the sinking of the Torrey Canyon only a few miles to the east of St Martin's on the Seven Stones Reef. Fortunately for the islands, the prevailing westerly winds carried most of the oil to more distant shores in Cornwall and finally to Brittany, so most of our own seabirds were saved though evidence of a decline in Guillemots followed the disaster, and other species may also have been affected. Much was learned from the results of the Torrey Canyon affair, and even more has been written so I do not plan to dwell on the subject here except to say that were a similar wreck to occur tomorrow, would we be any better prepared to combat the effects of pollution? I personally have my doubts.

One positive thing came of the Torrey Canyon disaster, and that was a survey of the breeding populations of Scilly's seabirds commissioned by the Nature Conservancy and conducted by John Parslow. I was able to assist in a minimal way in this by contributing counts from Men-a-Vaur and St Helen's. I also got to know John Parslow better during that period, being able to offer him hospitality on the floor of our Tresco cottage! He too was struggling a bit in those days, being to all intents a free-lance ornithological troubleshooter. He had no permanent job, but I envied him being able to work in an area to which he was deeply committed, though it never occurred to me then that one day it might be possible for me to do the same. Since those days John has gone on to become Director of Conservation for the RSPB and author of many successful books and papers on ornithological subjects, living proof that dedication is ultimately rewarded. And he is still a friend of mine, or ought to be after that effusive build up. He is also still very much in touch with Scilly I am glad to say.

Once our concern following the Torrey Canyon had been to some extent allayed, the remainder of that summer went by with little else to remember except that my parents came to visit in June, but before that planned event came an unscheduled surprise! This was the totally unexpected appearance one day of Baggers, his wife Pat and children. He was probably as amazed to see me as I was to see him, but it was a pleasant surprise, and I was happy to be able to show him some of the seabird life of St Helen's, where he did some filming which I believe eventually went out on Anglia TV. Dick had hardly changed during the twenty or so years since I had known him at Gresham's. His hair had whitened considerably, but apart from that he still had his boyish enthusiasm, and was just like the man I had so much admired and wished to emulate as a schoolboy! Although we had met briefly while I was gardening at Derry and Toms, it was a memory I had scarcely retained, probably because I had been embarrassed by that earlier encounter.

While Father was staying I took some time off to show him around a bit as well, and it was during his visit that I saw a bird I was convinced was a Gull-billed Tern though in retrospect I can't imagine why! It flew overhead calling raucously – a cry I did not recognise at the time, and cannot recall properly now. As it flew away and out of sight along the Tresco shore it seemed paler and bulkier than the Common Terns I was familiar with, but it did not seem to be a Sandwich or a Roseate Tern of which I also thought I

50. Settled in at the Blockhouse

had enough experience. Therefore what could it be? Answer: Gull-billed Tern of course! What a dreadful way to arrive at an identification; but having made it I did not want to go back on it, especially since Father was nearby though he did not claim to have seen it as well as I did. So when the time came to write up the sighting for the BBRC I sent them what I thought they would want to read, to substantiate my claim. Ironically, after all those rejections of honest sightings, on the first occasion that I perverted the evidence my record had been accepted without question! Luckily I did not continue with further deviations (with one exception which I will be going into later), for most of the time there was no need, as the birds concerned were genuine. However, I did learn one thing from this – that it was not difficult to convince the BBRC once one's credibility had been accepted by them as mine obviously had. I should say at this point that in 1983 the record was withdrawn at my request, thus salving a conscience that had been pricking me ever since the event! What do I think it really was? Probably a Roseate Tern which had moulted its long outer-tail feathers prematurely, as noted by Victor Tucker (*British Birds* 73: 264).

At home and at work, things continued much as they had the previous summer. I was still playing hide-and-seek with the Wing-Commander, while at the same time managing to do just enough work to keep out of trouble from the Commander (I hope you have grasped the distinction between the two by now). Actually, as I had found out, so long as the Commander was satisfied with the look of the hotel garden this was sufficient to keep George Leatherbarrow off my back, and apart from one or two minor differences of opinion, George and I had learned how to avoid conflict with one another to our mutual convenience.

Post-breeding Dispersal

51. The credible Long-billed Dowitcher

As the boys grew up, they were becoming more independent and Marianne was less concerned for their safety though I felt that, of us all, she was the one who had suffered most from the move away from London as Tresco had little to offer her apart from sunshine, which she loved. Her mother also managed to get over to visit us from Germany which was nice for them both, but for most of the time I'm afraid I neglected her in a shameful way, thinking perhaps that with the welfare of the children to look after that was enough to keep her happy. We made a few friends among the other young couples with children who like us had settled on Tresco in order to enjoy the 'good life' but had little in common with them apart from our isolation.

Derek Upton and his wife Audrey were among our handful of friends on Tresco, and together Derek and I dreamed up the idea of starting a mail order business, selling rooted cuttings of some of the easily grown succulents and other exotic plants which proliferated all over the island. Derek convinced me that it was a worthwhile idea, and I agreed to put up a very small amount of capital for the project. The capital was needed for press advertising, but first we had to get some stock, so we helped ourselves to a few hundred cuttings of the Hottentot Fig (*Carpobrotus edulis*) which grew in a wild state right outside the cottage, and planted them in a plot in Derek's back garden. Being the Abbey Gardens propagator he could not risk doing this at work for fear of being accused of robbing the gardens; he knew he was skating on thin ice but, nevertheless, was prepared to take the risk. Anyway, having done this, we sat back and waited for the cuttings to take root.

Meanwhile I carried on with my painting, and even found that I had one or two steady patrons – regular guests at the hotel who liked my work, and I suppose of its kind it was not bad. Among these patrons were Sir Frederick Watson-Jones, a Harley Street consultant and Robin Bailey, in those days a little known TV actor. It was pleasant to be able to chat with people of obvious standing, and even to be able to join them for drinks at the hotel bar in the evening though I could tell that poor old George Leatherbarrow was not too keen on the gardener hob-nobbing with his esteemed guests! Nevertheless he had to grin and bear it because in the end pleasing the guests was his first responsibility.

That autumn I consolidated my position as the leading bird person on Tresco (which was not difficult) but, more importantly, had gained more credibility with the BBRC after finding a Long-billed Dowitcher, later seen and observed by many other observers as it lingered on in the area of the Tresco Pools until winter.

I also discovered another Baird's Sandpiper on Bryher, yet another White-rumped Sandpiper on Tresco, and at last caught up with my first Buff-breasted Sandpiper on St Mary's – I was beginning to get around as well! The latter I found in company with Peter Mackenzie, who had also finally overcome his natural diffidence and accepted me as a fellow enthusiast. Until then I had looked upon St Mary's as just the island one went to to do a bit of shopping, or from where one caught the boat or helicopter which took one to the mainland. In common with most people living on an off-island I had the opinion that St Mary's was not as nice as where I lived, being too large, too populated, too noisy and too anything else one did not like! But as I slowly got to know people on St Mary's I began to realise that it had much to recommend it in terms of social contact. The larger population of St Mary's was in fact a much healthier community to live in I decided – on Tresco one was very isolated and inevitably often thrown together with people one had nothing in common with except the need for companionship. How nice it would be to be able to find a job on St Mary's and get away from the pressures of life on Tresco which I had originally gone to in order to escape the quite different pressures of life in London. But these were just pipe dreams and I knew it, and I reconciled myself to the advantages that Tresco did have to offer.

Winter activities on Tresco included annual shooting parties, when family and friends, often landowners and titled people known to the Dorrien-Smiths, descended on Tresco for a few days' 'sport' – which meant blasting off shotguns at the variety of 'game' available. The birds at risk were largely reared Pheasants, ducks both reared and wild of various species, Snipe and Woodcock, the latter which often arrived in considerable numbers in November and were considered particularly good sport. Although I disagreed with this on principle, I usually found myself volunteering to be a 'beater' on these occasions, partly because it gave me the chance to get out and about in areas which I otherwise rarely got to in winter, but mainly because it offered a change from the normal routine which could become pretty boring. Usually the Abbey gardeners would be conscripted, together with a few other estate workers and young or female members of the Dorrien-Smith family and friends.

These occasions were usually quite informal but it was necessary to 'know one's place' – for example at lunchtime large containers of rabbit stew would be provided for all, but the beaters, other than those in the family were required to sit at a respectful distance from the 'gentry' who, as far as we could judge, were served exactly the same food. When all his retinue had had their fill, the Commander would call across to the beaters and enquire if they had had enough to eat, and if requested the left-overs would be divided among us – though there were seldom requests for more.

Beaters, and there were no exceptions for family, age or gender, were all togged out in canvas smocks and leggings and expected to flog through the cover, often thick gorse, brambles and bracken, whacking the undergrowth with a stick and shouting appropriate arousing cries. This behaviour usually resulted in alarmed birds, game species or not, flying out in all directions in varying states of alarm. The gents with the guns would be either standing in advantageous places outside the cover, or abreast of the beaters, and their job was to recognise the species they were allowed by law to shoot and then try and hit them before they were out of range. At the same time the guns were expected to avoid non-game species and beaters when it came to pulling the trigger! This did not always happen, and although only one beater ever got shot during my time on Tresco there were a few near misses and several unfortunate owls and water-rails were shot in mistake for Woodcock.

Sometimes the gents with guns were lacking in experience, as exemplified by the chap who aimed at a pheasant that was craftily sneaking off into the undergrowth without taking wing.

'Don't shoot it while it's running you fool,' shouted the enraged Commander. (It was only 'sporting' to shoot at a flying bird.)

'I'm waiting for it to stand still,' came the reply.

Even the Commander was speechless at that!

On another occasion it was the Commander himself who nearly slipped up. We were driving Castle Down at the north end of Tresco and moving along line-abreast through the short heathland when quite a large bird flew over from the direction of the open sea. Usually quite a good shot, the Commander raised his gun and loosed off both barrels at the unidentified flying object. I was happy to see it continue flying, apparently unscathed, because I had identified it. I could not let the incident by without saying something.

'Why did you try to shoot a Great Northern Diver?' I found myself asking.

'Oh, I'm sorry, thought it was an old Cormorant,' was the Commander's reply. There was no answer to that either.

On another occasion we were working through some dense rhododendron thickets, a favourite lying-up place for Woodcock by day. A few birds had come out, but none had gone the way of the guns. We had a couple of youngsters among the beaters who had not been on a shoot before, and they had been giving vent to their high spirits with war-whoops and other cries not normally used to flush Woodcock or Pheasants. There were a few standard ritual noises that beaters were supposed to make, but these boys were unaware of them. After the drive was over the Commander called the beaters over to where he was standing.

'I've been hearing a lot of Cowboy and Indian noises from some of you beaters,' he said. 'That isn't what we want. The correct way to flush Woodcock is rrrrrrrrrrrrrrrrRRRRRRRRRR My Cock, My COCKY,' he demonstrated as if born to it. I wonder if he really would have had better luck if those beaters had used the proper call? Anyway, after stifling back our giggles at the Commander's

52. *'Ah Hunt – conserving your energy I see'*

effort to explain the finer points of the art of woodcock flushing we went back into the thick of it.

And thick it often was, sometimes painfully so in the gorse of St Martin's where the shooting parties occasionally went as an alternative to Tresco. I remember seeing some small *Sylvia*-type warblers in the gorse there one winter which I never certainly identified. I thought at the time they were Lesser Whitethroats, but now realise they may have been Dartford Warblers, which are particularly fond of gorse.

As beaters we did not enjoy the gorse too much though, especially on St Martin's where there are literally acres of it. On one occasion I was taking the least line of resistance, trying to avoid the thicker stuff, when the Commander spotted me.

'Get in there Hunt, what's the matter with you?' he bellowed.

'Trying to conserve my energy, Sir,' I replied. He never let me forget that remark, which he thought highly amusing. If ever he caught me taking it easy in any circumstances he would always come out with it.

'Ah, Hunt – conserving your energy I see, Ha Ha!' Actually although I fell out with him on several occasions, especially later on during my days on Tresco, I still respected the man, and in a peculiar sort of way I quite liked him despite his autocratic ways. Not that my behaviour on Tresco was blameless – far from it in fact, and if he had known about some of the things I got up to I would probably have left Tresco sooner than I did.

I even had one of his Pheasants on one occasion, though the Pheasant was in part to blame for that. It was all because of its fondness for my Brussels sprouts which it kept on stealing. One day I found it entangled in some garden netting and applied the final solution – I wrung its neck! This may sound unkind coming from a supposed bird-lover, but if there is one thing I like better than Brussels sprouts it is roast Pheasant. Luckily there were some sprouts still left, and we were able to enjoy them both together.

Then there was the occasion of the Christmas tree. Unlike on the mainland where it is possible to buy a tree in every corner shop or supermarket, in Scilly it is not so easy, and on Tresco in the 1960s it was not only difficult but expensive to buy an imported one on a gardener's wages. In any case, surrounded by potential Christmas trees as we were, it seemed ridiculous actually to buy one! The main snag was that if everyone decided to take the top out of a growing tree ever year there would soon be no growing trees left, so it was discouraged to say the least. However, I decided to help myself and together with a friend planned a daring strategy for securing a tree in broad daylight. Knowing that the Commander was in the habit of attending church every Sunday at eleven, together with most of the island folk who might possibly inform on us (and there were some), we planned to meet in the woods at about eleven-thirty to perpetrate the great tree robbery. The main problem was choosing a suitable tree. We did not want to desecrate anything, in fact both being gardeners we were very conscious that we were doing wrong and therefore anxious to remove a growing point that would make a good tree without spoiling what we left behind. So it took us longer to make up our minds than we had allowed, and we were just in the process of carefully sawing out the chosen tip when we heard voices!

'It's the Commander,' whispered my friend, petrified. And he

53. *A Wryneck at last*

was right. What's more he was looking straight in our direction and gesturing with his stick! We froze to the tree-trunk – we were about thirty feet up as it happened, and waited for the inevitable. To our astonishment nothing happened, and when we dared to look again at where the Commander had been there was no sign of him. Our luck was obviously with us that day, so we finished what we had started and hurried home with the spoils. Needless to say I never took another Christmas tree while living on Tresco.

In 1968 I at last caught up with that elusive Wryneck, or to be more accurate it caught up with me. Unusually for Scilly it was a spring record, and the bird frequented stone walls in the lane less than fifty yards from our cottage, so I was able to watch it at some length and at close range as it systematically fed on ants which were nesting in loose soil among the stones of the hedge. I could even see the rapid movements of its tongue darting to and fro as it picked them off one by one. Another exciting spring visitor that year was a Woodchat Shrike – although I had seen a few drab juveniles in autumn before, this was my first adult in breeding plumage and it made a perfect picture perched in the gorse, which was in full flower at the time.

This was also the year when 'Seabird Special' boat-trips got under way. Boatmen Vernon and Michael Hicks were running the *Sea-horse* as an independent venture for holidaymakers staying on Tresco, and I teamed up with them to do a once-weekly special trip among the nearby islands, looking at the various seabirds and shorebirds. This proved to be instantly popular, and since it brought in a couple of quid extra per week was worth doing from my point of view, and it also gave me the opportunity to visit places I did not dare to go to in my own punt.

News of this success got through to St Mary's, and I was first approached by members of the St Mary's Boatmen's Association; but they could not agree about using an outsider, deciding that if bird-trips were required they could do them without me. Then along came Cyril and Garfield Ellis, joint owners of the independent launch *Buccaneer*, and we started regular Sunday trips which are still going!

At the same time I had started giving talks about birds to visitors in the Tresco school, illustrating them with my own tape-recordings of bird-song and slides which had been loaned me by the RSPB. Following a request from that organisation for volunteer representatives I had offered my services, and was promoting the Society at the same time as my boat trips! Looking back on it I realise I must have been incredibly busy then – no wonder my boss had the idea I was not giving the job my full attention. What Marianne thought of my neglect of her and the children I was to find out in due course, although at the time I was quite unaware of how all this was affecting our marriage, being too immersed in my new-found interests. I suppose I thought that because my various activities all contributed to our family budget in one way or another, that was justification in its own right – if I thought at all that is; now I have my doubts. So when I discovered that Marianne was seeing rather more than I would have liked of one of our neighbours, who was also one of my work-mates, I was not only shocked but surprised! I could not understand how she could prefer his company to mine, when if the truth be told she was seldom in my company except when I was snoring beside her in bed. But I was neither prepared, nor could I afford, to give up all my extra activities which by then were earning me almost as much as I was getting from my pay-packet (which had risen to £12). Naïvely, I thought that now I had found out about her little romance and shown my disapproval that would nip it in the bud, and I carried on with the boat-trips, the painting and the lectures just as before. I even bought myself a proper camera and a telephoto lens, because the borrowed slides could not illustrate local scenes or bird life, and besides that I knew I could do better myself!

One of the hotel guests I met that year, Martin King, put quite a novel proposition to me.

'I'll pay you £1 for every bird you can show me on this list,' he offered. The list included Manx Shearwater, Ring Ouzel and a few more I cannot recall, but the only possible one, because it was not during the migration season was Manx Shearwater.

Post-breeding Dispersal

54. As photographed by Martin King

'Forget the money, but come to the New Inn just before closing time tonight, buy me a drink and I'll show you some shearwaters,' I replied, knowing that most nights they were in the habit of coming ashore at the north end of Tresco, where a few pairs probably nested.

It was a damp, drizzly sort of night, just right for shearwaters I thought, but when I got to the pub there was a message from Mr King to say that since it was not a nice night he would have to call the expedition off! No wonder he had never seen a Manx Shearwater I thought, but what made me curious was that a chap was prepared to pay for a bird but not suffer for one. For me at least, part of the fun of a new bird is sometimes the agony one endures in getting to see it – perhaps the pain for him was to part with the money! This turned out to be a likely solution, as I later found out that Martin was a bank manager. I did eventually get him to part with some cash – but not for a bird. He bought one of my paintings!

It would be true to say that my encounter with Martin King marked a turning point in my life, for it was he who awakened me to the fact that people existed who were willing to pay me to show them birds. He also sent me a few brochures advertising tours in Scotland with naturalist leaders, pointing out that I could be doing the same sort of thing in Scilly. It was food for thought.

Through the regular Sunday trip in *Buccaneer*, I got to meet Richard Coomber who was then working in Lloyd's Bank on St Mary's. He was also a very enthusiastic bird-photographer – he helped me to promote the trips from St Mary's where unknown to me a boating conflict was beginning. The *Buccaneer* at that time was very much an independent outfit, struggling for business against the combined might of about a dozen launches operating together as the Boatman's Association. Stung by the success of our 'Seabird Special' in *Buccaneer* which, remember, the Association originally had turned down, they were forced to operate a rival trip with skipper Lloyd Hicks, himself formerly an independent in *Swordfish II*. Lloyd was not only an old sea-dog, but a knowledgeable chap with a keen interest in birds and a good talker. He gave a regular slide-show on St Mary's featuring his own photographs, and was a formidable rival from the start, though living on Tresco I was blissfully unaware of all the goings on on St Mary's – all I had to do was wait for the *Buccaneer* to collect me at Tresco Quay, give my commentary as the boat cruised among the seabird colonies, collect my share of the proceeds, and get off again at Tresco.

Meanwhile my little mail-order venture with Derek Upton was also beginning to bear fruit. Calling ourselves Scillonia Horticultural we had placed an advertisement in one of the gardening journals, in which we offered rooted cuttings of what I think we called 'Giant Flowering Mesembryanthemums': at three for £1 it was money for old rope while it lasted. Unfortunately, one of the orders got put into the wrong mail-bag and ended up on the Commander's desk in Tresco Abbey! Poor Derek – being the Abbey Gardens' propagator he was accused of attempting to sell the entire stock from under the Commander's feet. He was required to leave the island in disgrace and bore the full brunt of the Commander's wrath – I was merely put on the carpet and warned that if I did not settle down to my job and pay less attention to my outside interests I might have to follow Derek and look for another job myself. I began to wonder how long I could continue anyway, but the problem was could I afford to pack up and go? Although we had been helped with our removal expenses in getting to Tresco, nobody would help us leave except with the toe of their boot! I began to realise just why the whole business of living in a tied cottage was considered so undesirable.

To make matters worse, Marianne was still having an affair which seemed to be getting more intense; there seemed to be nothing I could do about it except grin and bear it, hoping that in time she would come to her senses (I could still not see that I was largely to blame). Luckily the autumn season of bird-watching came along at that point and I was able to immerse myself in that. As it turned out, 1968 was one of the greatest years on record for American vagrants and I had a good share in the action.

But before the arrival of the Yankees, Britain was invaded by a host of birds from quite the opposite direction, some of them even penetrating as far as Tresco. As I was cycling to work after lunch one afternoon in September I saw a bird fly in from the direction of the beach and settle for a short while on a clothes-line pole only about twenty yards away. I at once recognised it as a Nutcracker but, as it flew off in the direction of a nearby pine-belt and I had to go off to work, I was unable to follow it up until the following day when it or another bird gave me a fine view among the pines. The Nutcracker fed by ripping bark from a dead and decaying tree, thus exposing earwigs and woodlice which it picked off eagerly, occasionally dropping to the ground to seize an insect which had escaped. As it fed it was followed by an opportunistic Robin which helped itself to whatever the Nutcracker missed! Although I had seen the species previously in Norway, that occasion was a distant memory, and in any case it is always much more exciting to find a new bird on one's own patch and get close-up views of it than to get distant views of a bird where one is expecting to see it. Later on I learned that the Nutcracker invasion had started on the East Coast in August, and that over 200 had been recorded in Britain during the autumn. They were considered to have originated not from Scandinavia, but from Russia, following a failure of their favourite food – seeds of the Arolla Pine.

A guest at the Island Hotel that year was none other than Ian Wallace, who spent a week there with his family in late September. Ian was one of the best-known field ornithologists and an established member of the BBRC at that time.

I was amazed by Ian's ability to spot and identify (and almost always correctly) small birds at a glance. I had always considered my own eye for a bird to be good, but Wallace was quite outstanding. I was also impressed by his skill at sketching what he saw in the field with just a few lines on a pad that he always carried with him – I had not met anyone apart from Richardson with this enviable ability, and it made my own wooden caricatures of the birds that I had the effrontery to be selling seem even poorer in my own eyes than they in fact were. I realised that I had an awful lot to learn about observing and painting birds, and then to find that despite his distinguished appearance and great experience of birds at home and abroad he was a fractionally younger man than me came as a great shock! What had I been doing all the time while Ian had been acquiring all that expertise? The answer of course has

Post-breeding Dispersal

55. Northern Waterthrush – definitely not a BBRC plot

already been given in the earlier chapters of this book. I obviously had to make up a lot of ground if I wished to be ranked among the better ornithologists of my generation.

One day Ian came back with the story of having heard a bird calling in the overgrown swampy vegetation by the Great Pool which he had only glimpsed, but was sure was one of the American species of waterthrush. It was essential, he said, that if possible I should get to see it and confirm his suspicions; better still, try to see enough of it to establish its exact identity. He told me the spot to watch in and what to listen for and left me to it. It was arranged that during the day we would take turns to wait and watch the area in hopes of the bird returning. This I did as Ian had directed but, after what seemed like hours, still nothing had turned up and I began to believe it was all an elaborate trick specially arranged by the BBRC to find out whether or not I was a reliable observer. Suppose I was to say I had seen the bird when in fact it did not exist – then they would know once and for all that I was not to be trusted! These paranoid thoughts, probably a result of my feelings of guilt relating to the Gull-billed Tern deception, were becoming more and more prevalent in my mind when I heard in the distance an unfamiliar call coming from the undergrowth. The sound came nearer and nearer. 'Chink...chink...chink...CHINK' as suddenly it was almost on top of me – then 'chink…chink…chink…' as it gradually faded away without my having seen a feather of the bird. It was a relief to know that it was not a BBRC plot, but tantalising to have been so near without seeing it. Of course I waited again and eventually got some brief but excellent views. Ian did as well, and in the end we saw enough of the bird to claim a positive identification of Northern Waterthrush, the second record for Scilly and Britain. We had to wait until 1982 for another to appear, this one equally elusive, but on the island of Bryher.

The excitement of autumn 1968 continued. One lunch time there was a knock on my door and a very breathless bird-watcher who I later learned was called Steve Joyner told me he had just found what he believed was an Upland Sandpiper in a ploughed field only a short distance from our cottage. Dropping my knife and fork (every bird-watcher's wife will recognise the syndrome) I replaced them with my binoculars and ran after Steve, who was already hastening back to where he had last seen the bird. It was not long before we had relocated it, this time in a grassy meadow. Unfortunately it took wing before I could see it properly on the ground, but as it flew it showed a longish barred tail, and uttered a very unfamiliar call. I too felt sure it was an Upland Sandpiper but, as neither of us had seen enough for a description to satisfy the BBRC, we knew we would have to find it again. We searched every likely field but with no success: as it had flown off strongly in the direction of St Mary's this was not too surprising. That evening I got in touch with Richard Coomber, who finally located it in a grazed field close to the hospital on Peninnis Head. Here it stayed and was seen by many observers, including several islanders, who were just starting to realise that bird-watching was not just a hobby for eccentric retired colonels and hearty maiden ladies. Eric Woodcock, who was farming on Peninnis in those days later told us that the Upland Sandpiper had first appeared in his field about an hour after Steve and I had watched it fly away from Tresco. Eric as I then found out was a neighbour of Peter Mackenzie, and himself a keen photographer with an interest in birds.

Meanwhile, Ian Wallace had moved over to St Agnes where a Red-eyed Vireo and a Blackpoll Warbler, both in those days extreme rarities were discovered. St Mary's too was coming under close scrutiny from bird-watchers, who were no longer confining their activities to St Agnes – in fact with a Bobolink, two dowitchers and later on in October a Myrtle Warbler discovered there, St Mary's had as much attention as anywhere. On Tresco, having to get on with the job at the hotel prevented me from seeing almost all the rarities being reported elsewhere, and I was dreadfully frustrated, especially since I could not find anything myself! I spent my first hour of 'work' every day in carefully checking out the garden and other areas adjacent to the hotel. As soon as

I finished I cycled furiously to the pools in the evening, but with little success, apart from a few routine migrants – and all the time news of fresh rarities on other islands kept getting back to me. Coupled with my worries at home and at work I became very depressed indeed.

At last, one misty morning in late October I heard a bird calling in some tamarisks close to the hotel which was unfamiliar, so definitely worth checking out. I spent the whole morning pursuing it through the damp undergrowth of bracken, brambles and overgrown hedges. Luckily, because it was a miserable day, my activities did not come to the notice of George Leatherbarrow, or perhaps by now he was so used to my shirking that he would have been surprised to find me at work! Anyway, after becoming thoroughly soaked in my efforts to identify the mystery bird, which only gave away its whereabouts by calling 'Chack – chack' persistently, I finally decided it must be a Dusky Warbler, though I badly needed someone to confirm it.

The previous evening I had also seen a bird which I had identified as a Rustic Bunting, though this was not considered such an extreme rarity in those days as the Dusky Warbler, a species only recorded once previously in Scilly and a handful of times in Britain. With these two birds as bait, I phoned St Agnes to try to persuade some observers to come and help me out with my identification problem. Ron Brown and Alan Greensmith were seemingly the only watchers there. I told them first of all about my probable Dusky Warbler. To my astonishment they didn't seem very interested.

'Saw the one at Dungeness last year,' came the reply. I was not used to this sort of reaction – I thought that any rare bird was worth seeing more than once, and if someone requested my help in identifying a bird I would have gone straight away if possible.

'There's a Rustic Bunting as well,' I had the good sense to continue.

'Oh – we'll come for that' – this apparently was new for them.

Next day I could not find the warbler and Ron and Alan failed to find the Rustic Bunting. I was glad they had failed and when I relocated the bunting later I didn't bother to let them know. My description of the Dusky Warbler was sufficient to get accepted by the BBRC and, having seen others since, I am sure my identification was correct. The Rustic Bunting too was accepted, and I later saw it or others about Tresco on several occasions.

I also managed to get over to St Mary's one afternoon to see the Myrtle Warbler, which ironically was discovered by Ron Brown and Alan Greensmith, later to become legendary characters in their worldwide pursuit of birds. But in 1968 they were not my favourites.

The excitement was still not over, for in November I was out looking for any late migrants that might still be lurking on Tresco when a large butterfly (so large that at first I thought it must be a bat) soared out from the top of a clump of pines and glided down to nearer eye-level. I was astonished to see the intense rufous wings with black veining which could only mean that it was a Monarch – a North American species which like the small birds can sometimes be carried across the Atlantic by fast moving weather systems at migration time. As I watched it a sparrow flew out from the trees and seized the Monarch in mid-flight; but after a few seconds the butterfly was free and flying rapidly away across the fields – the sparrow had discovered that the Monarch is protected from predators by being extremely distasteful. Presumably birds in North America have a built in awareness of this, but being a Scilly sparrow, this one had had to find out by experience!

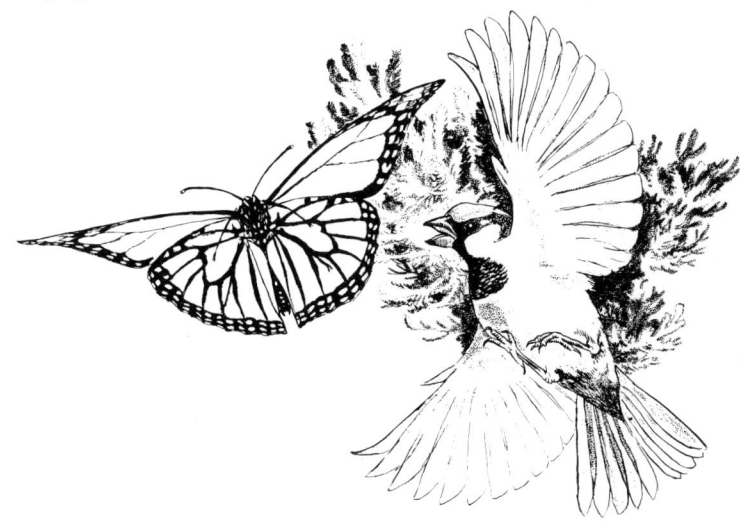

56. *'Sorry – my mistake!'*

Because I was concerned about our marital problems I made every effort to get away from Tresco, including writing to all my relations, well-off patrons and others in an attempt to borrow enough money to put down a deposit on a small guest house on St Mary's. I was convinced that I could make a go of running an establishment catering specially for bird-watchers and other naturalists. This idea stemmed partly from Martin King's suggestion that there were those who would willingly pay me to show them the wildlife of Scilly, and also because I realised that I could earn as much from two weekly boat-trips as I was being paid for a week's work in the hotel garden. Together with my earnings from painting, I felt sure that I could soon pay back any borrowed money with the guest house takings. I even got so far as to look over a house that was soon coming on to the market – I decided to call it the Shearwater Guest House, and wrote a brochure for the project which I sent around to all potential sponsors. I received a few replies and a fair amount of sympathy, but no money. The house was sold to someone else, who hit on the idea of calling it the Shearwater Guest House, and this it remains.

Finally, in the spring of 1969 I decided that the only way out of the dilemma was to take drastic action; I could no longer stand the strain of doing a job I no longer got any satisfaction out of while my wife was having an affair – so I gave in my notice. This I figured would solve the problem, and it eventually did, but not without a great deal of unhappiness on all sides. It is a period in my life I have largely managed to expunge from my memory. It makes me feel bad as I write it but, for the sake of continuity in my story and because these after all are 'confessions', here goes.

When the time came for us to leave the cottage, Marianne and the boys went off to stay with her mother in Germany – not because she wanted to but because there was no option. For my part I planned to stay with a family on St Mary's as their lodger until I could find a place for us all to live together again. I had an assured small income from the boat-trips in *Buccaneer*, with day excursions about the islands which I had planned, and I also had a part-time job at the Star Castle Hotel as a gardener! Plus of course additional money which I would be able to get from selling paintings – I had a well-stocked portfolio of both landscapes and bird pictures that I had mass-produced during the winter. I was quite confident of success.

6. The Adult Birdman Emerges

Luckily I had met Sue and Andy Brooks (Andy worked as farm labourer for Peter Mackenzie) the previous summer and they had kindly offered me the use of their spare room while I was searching for accommodation for my own family. I doubt if they would have been so generous if they had known how long it was going to take, for their cottage at Number 6, North Parade became my headquarters for the next two summers. I was even allowed to put a notice-board in their front garden advertising my various activities which soon included, besides my boat-trips and 'safaris' as I decided to call my other excursions, a regular Saturday evening slide-show to rival the Sunday one by Lloyd Hicks. Peter Mackenzie, however, had noticeably cooled towards me – he obviously disapproved of my having taken such drastic action, and also I suspect did not welcome another fish in his pond. Peter was well in with the Duchy of Cornwall; besides being the local vet, he was also heavily involved with the newly established Museum Association which had recently opened in splendid new premises. My activities, even though I was recognised as RSPB representative, were generally not approved of by the establishment on St Mary's and Mackenzie was very much an establishment figure. So from that direction I was disappointed to find only very guarded enthusiasm – in retrospect I do not find this surprising, but at the time I was quite hurt. Nevertheless it did nothing to deter me from my intention of becoming the top-dog in Scilly ornithology. Richard Coomber continued to help me in many ways, and being the only bird-watcher close to my own age and therefore better able to understand my problems, he was a valuable ally.

That summer I gradually found out what I had let myself in for! The boating rivalry had escalated into near warfare, and inadvertently I had become a front-line skirmisher on the side of the inde-

57. Spot the real Buccaneers

pendent *Buccaneer* when all I wanted was to be at peace with the rest of the world while fighting to save my family from splitting up irretrievably. Nevertheless if I was to succeed in this I would have to be able to offer some financial backing as well as a roof over their heads, and it was clear that since I had thrown in my lot with the *Buccaneer* it was with them that I would have to earn my daily bread – the Association boatmen would not be likely to offer me any concessions now that I was clearly affiliated to their rivals.

The Sunday 'Seabird Special' was the first line of attack because it enabled us to get a crowd of interested visitors on the *Buccaneer* the first day after their arrival on holiday (most of them arrived on Saturday). So if I could get a crowd of people to my Saturday slide-show I then had to promote my Sunday trip heavily, and once they were hooked on the *Buccaneer* it was up to Cyril and Garfield to keep them as regular customers for the rest of the week. Meanwhile I did all I could to promote my 'safaris' to Tresco and St Agnes, giving a few extra customers to the boat, and putting a few extra pounds in my pocket at the same time. The other boatmen naturally did eveything they could to stop visitors going with *Buccaneer* – although it fell short of physical violence there were times when it very nearly erupted. The usual tactics were not very subtle, ranging from the 'disappearance' of our notice-boards and leaflets to the expression of grave doubts as to the seaworthiness of the *Buccaneer*. Visitors in all innocence enquiring for David Hunt at the quay were answered by 'Never heard of him' or 'You mean the kitchen porter from Tresco?' And if they asked for the 'Seabird Special' would be directed hastily in the direction of the ticket office where they would be sold a ticket for *Swordfish II*. It was all perfectly fair and above-board. In spite of the efforts of the other

58. Sunday 'Seabird Special'

boatmen, enough customers filtered through on to the *Buccaneer* to make our trips worthwhile. In fact the tactics used by the others often had the effect of sending discerning visitors our way – especially the notice which went up one day on the Association's board: It read – 'ALL ASSOCIATION SKIPPERS ARE FIRST CLASS ORTHI-NOLOGISTS'. Of course we did not point out the mistake – it just helped us to fill the *Buccaneer* and keep smiling! Of course we lost a lot of potential customers as well – people on holiday are not anxious to get involved in local politics or take sides in personality struggles. Regular visitors were naturally afraid of offending the boatmen who they regarded as friends, and once an allegiance to a particular boat or boatman had been formed it was difficult to patronise what was clearly a rival outfit without jeopardising a valued friendship. So we lost some as we gained some, but over the next few years we slowly built up a regular clientele despite anything the twelve-strong fleet of Association skippers could do.

Steadily my safari trade increased as well, but it was very much influenced by the weather. I realised that it was not an income I could rely on. I was managing to earn a bit from painting too, but to my dismay I found that my part-time gardening job was the only dependable one! The owner of the Star Castle then was an eccentric gentleman called Roland Stephenson, who was also the local practising solicitor. Luckily for me he was sympathethic to my various problems, and willing to employ me when I was available. In practice, this was more often than I would have liked but at the same time I was glad of the money I could earn at the Star Castle to subsidise my other meagre income. My main duties were to keep an eye on the greenhouse and vegetable garden, and give instructions to an inexperienced full-time man who was responsible for the routine maintenance – just the sort of job I should have liked on Tresco!

The Adult Birdman Emerges

Ironically it was to Tresco that I still had to turn for a fair amount of my business. I continued to do my slide-shows in the school there on Sunday nights, and led a safari there the following day. After the evening show I would go up to the hotel with my portfolio and attempt to sell a few paintings. This required a fair amount of nerve, for although I had not left the Island Hotel in the lurch (I think George was rather glad to see the back of me), I had left with rather a cloud hanging over me, and felt ill at ease going back. But the hotel guests were likely clients for both safaris and the sale of paintings, so it had to be done.

The first Monday evening I went into the hotel bar and found George himself behind the counter as it was the regular barman's night off. He put on a show of making me welcome, but I knew he was unhappy to serve me. Next week he was nowhere to be seen on either side of the bar, which was very unusual. I asked Duncan, the barman, what had happened to his boss.

'He asked me to change my night off so he wouldn't have the embarrassment of having to serve you,' came the reply.

I had never realised before just how my attitude must have affected poor old George Leatherbarrow – it seems our animosity had been mutual, and I just thought it had been my feeling for him!

In retrospect I realise that George's attitude towards me had been very much prompted by worries of his own, especially since as hotel manager he had been answerable to the Commander for the state of the garden, and if I was not around to answer why things were being neglected (and make no mistake they were), then it was poor George who got it in the neck. Knowing that he knew a lot less about the subject of gardening than I did put him at a disadvantage which I always exploited if given an opportunity. The added irritation of having a gardener who hobnobbed with the guests as equals, and who was also knowledgeable about many other subjects outside of the job must have contributed further to his dislike.

A few years ago George retired and came to live on St Mary's and I have found that we now get along very well, although of course we do not see a lot of each other in the normal course of events.

Garfield Ellis, co-owner of the *Buccaneer*, helped me to shift what personal effects were worth salvaging from the cottage at Blockhouse and one fine day brought the *Buccaneer* to Tresco quay where we loaded everything on board. Once on St Mary's, I got permission from Barry Mumford who had some empty farm buildings at Trenoweth, to let me store our meagre possessions there until such time as I would need them again – which I fervently hoped would be quite soon.

Garfield and Cyril, who most people thought were brothers, were in fact cousins and made an excellent combination. They had an engaging way with the visitors; once charmed by either of them,

The Adult Birdman Emerges

59. *Cyril Ellis*

or both, it was very difficult for holiday-makers to go on one of the other boats. I too found them very friendly and helpful, and my allegiance was soon won, even if I had not been in partnership with them.

I kept in touch with Marianne by letter, but there was little in her replies to give me much comfort apart from the knowledge that they were alive and well. Even then it was a shock to hear that she was planning to set up home with her boyfriend in Cornwall! I could still not believe she could prefer him to me, and I was quite certain how Nicholas and Martin felt about it. I had of course precipitated the affair and now had no way of holding it in check.

They found work at a hotel in St Ives where Marianne was appointed housekeeper; as is usual in the catering industry, they were offered sub-standard accommodation tied to the job. As far as I was concerned this was the last straw, but until I could offer her a home that was better I could not expect Marianne to stay at home with her mother indefinitely while waiting for a call from me that might never come.

I paid my parents a visit to try and explain what had happened, but they could offer little but sympathy – I'm sure they appreciated Marianne's point of view better than I did in any case. I did not stay long at Newton Ferrers – there was nothing to do but feel sorry for myself, so I hurried back to St Mary's where at least I had found myself a social life. The Brooks family with whom I was staying consisted of Andy, his wife Sue and two boys a little younger than my own called Miles and Martin. I felt almost part of the family while I lived there, and it was a big help to have someone to talk to about my problems – luckily Sue was a sympathetic listener, and although I did not take much notice of her advice on how to deal with my marital traumas I owe her a lot for putting up with me for as long as she did. Andy was an easygoing chap most of the time, though given to occasional (justifiable) outbursts of impatience, usually directed at Sue or the children. He had a lot to put up with in that household, including my presence, but for the most part he stood it remarkably well. Andy was a keen jazz enthusiast, and had assembled quite a collection of records embracing a variety of styles both ancient and modern – in that respect he was very much a kindred spirit, and it was perhaps listening to his records that kindled in me once again the desire to perform.

While I had been on the mainland I had found a battered but playable trombone hanging in the window of a second-hand shop; it had only cost eight quid which, in the end, turned out to have been money well spent because it not only gave me an interest far-removed from my boat-trips and slide-shows but also gave me an identity much easier for island inhabitants, hotel workers and visitors to understand than the rather remote and specialised image projected by the artist-naturalist.

After only a couple of weeks' practice I felt willing and able to join Fred on his accordion, sometimes augmented by washboard, spoons, guitars or whatever in spirited renderings of jazz standards, Cornish folk-songs, 'Good-night Irene' or whatever else was required by the discriminating audience which used to gather every evening an hour or so before closing time in the Mermaid Inn close to St Mary's Quay. In return for this instrumental wizardry I was fortified with pints of ale and the knowledge that the great Scilly public was delighted by my virtuosity – there cannot after all be many artist-naturalists around that can play Tiger Rag as a trombone solo (come to think of it I don't believe there are any), and certainly no trombonists who can tell a Shag from a Cormorant! Except perhaps Roy Williams.

My first autumn of freedom from the serfdom of Tresco gave me plenty of opportunity to chase rare birds, though as it happened not a lot turned up. Even Tresco's Pools did not live up to what I had come to expect from them. There was one bird though which did cause something of a stir, and the ripples have not yet subsided.

The bird in question was a stint which after close inspection I

60. *Semi-western Sandpiper?*

concluded must be a Semi-palmated Sandpiper. Richard Coomber, Peter Mackenzie and others were persuaded to agree with me, and Richard was able to get some photographs. The accumulated evidence was then submitted to the BBRC for consideration and duly accepted. End of story you might think, but a few years later the record was reassessed in the light of new criteria by Ian Wallace in a paper specially prepared for *British Birds* (72: 264). The record was found unacceptable as a Semi-palmated Sandpiper, reflecting an opinion I had also arrived at in the interim period. My reasons for changing my mind about the identification are quite simple, though not so easily explained.

Firstly, at the time of the sighting (19 August 1969) the differences between two very similar North American species, Semi-palmated and Western Sandpipers, were neither understood nor even vaguely documented in the literature available. In fact, the Western had only once been recorded in Britain, on Fair Isle in 1956 (and first recorded as a Semi-palmated). At the time I had considered Western, but ruled it out almost at once because it was supposed to have rufous in the scapulars (the bird did not) and a comparatively long decurved bill. Although the Tresco bird had a longer bill than the average stint, the choice between the two seemed to favour the Semi-palmated which of the two was also, I thought at the time, far more likely as several had been reported and accepted by the BBRC during the last few years in Britain, making one in Scilly somewhat overdue.

So it was with a strong bias towards the Semi-palmated Sandpiper that my report to the BBRC was submitted. Even so, in the light of more recent information, from all the details of my description and Richard's photographs the evidence points strongly towards the Western, but with one annoying anomaly – my transcription of the call which in my stupidity I rendered 'chip' or 'cherp', the latter a conscious 'bending' of the truth in order to fit with my firm conviction at the time that the bird was a Semi-palmated.

Returning to Ian Wallace's paper now, his stated view is as follows: 'The balance of opinions within the Committee (BBRC) is that it was a short-billed Western Sandpiper, but as transcribed the call points to Semi-palmated. Sadly it must remain indeterminate.'

Had I been content with the truth and merely rendered the bird's call as 'chip' alone, it could now be accepted as a firm record of Western Sandpiper, and the first for Scilly. Which all goes to prove there is many a twerp between chip and cherp!

Bill Oddie in *BOLBBB* (*Bill Oddie's Little Black Bird Book*) explains how to fool the Rarities Committee, and good advice it is too if you need that sort of thing, but I would like his advice now on how to convince the Rarities Committee you are now telling the truth after admitting to having fooled them. Over to you Bill!

It has taken me years of consideration wondering how to relieve myself of that burden, and having done it I feel quite purged – now you know why I called this book my confessions!

I was sorry to see the departure of Richard Coomber in October of 1969; he had been a loyal ally in my struggle for recognition on St Mary's, and a good bird-watching companion. But Lloyd's Bank had other things in mind for him, and promotion meant being sent away from Scilly to a mainland branch. As it happened, Richard had other things in mind for himself, and it was not long before he was following my example, though on a firmer financial foundation by settling on the Hebridean island of Mull where he now runs a guest house and conducts similar style excursions to my own.

It continues to amaze me that, with the huge popularity of bird-watching today, more ornithologically minded people have not come to settle on Scilly. Perhaps it is the high cost of property, perhaps the isolation, perhaps they feel that one resident birdman is enough (some might say too many). Whatever the reason, I am still the only mainlander (you do not qualify to be called a Scillonian unless your parents were born here) to be seriously interested in birds to have settled here.

And there are only two Scillonians: Ron Symons, who works on the quay, and is an excellent observer of all things natural, but especially the seabirds of the smaller islands; and Francis Hicks, a young farmer on St Agnes, who grew up there during the heyday of the bird observatory, becoming a qualified ringer while still a schoolboy. Although he does not have as much time for bird-watching as he might like, Francis keeps his eye on things during the course of his work on the farm, and is often first to spot an unusual migrant.

When the 1969 season came to an end I was no nearer to finding accommodation for the family than I had been at the beginning. I had a lot of other things to worry about and having found a tiny chalet for myself to live in during the winter I looked around for a new means of financial support. Barry Mumford was the largest flower farmer on St Mary's and I approached him about the possibility of a job picking flowers as soon as the season commenced. He was quite willing to offer me a job but pointed out that there would not be any work until the New Year when the main crops came in. Until then I had no choice but apply for unemployment benefit, as I had not been able to save much during the summer even though I had earned enough for a good time!

That Christmas I managed to persuade Marianne to bring the boys to Newton Ferrers, and it was good to see them again. She was obviously looking after them much better than I had been looking after myself, though I put on a bold show of prosperity and claimed to be making a success of my independent venture. However, I was not able to offer her the one thing that I felt we really needed – somewhere where we could all live together again. I was still sure that if only we could once more settle down under one roof our difficulties would be over, I could get on with earning a living in the various ways I now knew were possible, and the children could enjoy a normal family life.

But when Christmas came to an end we went our separate ways once more – Marianne went back to St Ives and I returned to the flower-picking job Barry had promised. It was hard work, as anyone who has attempted it will affirm. Although I had helped out a bit on Tresco, it had only been a diversion rather like bush-beating, and I soon found out what a tedious, backaching job flower-picking can be. Luckily though, I was in congenial company, and Barry did not mind his workers having a bit of fun as long as the flowers were picked. It was also an open-air job, to which I had become fully accustomed, and provided one had reasonably waterproof clothing it was a lot pleasanter than I would have found working in a factory or an office. Finally I knew that when the picking season ended I would be able to resume my programme of boat-trips and safaris, which I had now worked out on a weekly basis, with a printed brochure and check-list of birds. I looked forward to the season with renewed optimism, even though there was still no prospect of a home for the family. I had to move out of my winter chalet to allow it to be let to holiday tenants, and was glad that I was still welcome back at North Parade with the Brooks family.

I had also been asked to contribute to an exhibition to be held in Plymouth of wildlife paintings by artists from Devon and Cornwall. This was a big boost to my confidence, though I realised that my work would have to be up to a better standard than what I had been churning out to meet the demand from holiday-makers in Scilly. I prepared several larger paintings of bird groups in island landscapes, put inflated prices on them and hoped for a few sales.

The Adult Birdman Emerges

61. First (and last) exhibition at Plymouth Athenaeum

The exhibition was well received and got quite favourable reviews, but I only sold a few pictures – it was still the pot-boilers at around a fiver a time that people wanted. In fact the only large paintings that did sell were to family and friends, anxious no doubt to boost my morale a bit. But I began to realise at this point that I was a birdman rather than an artist, and would be better off to concentrate on what I did best in future – boat-trips and safaris.

Spring that year was as exciting as an average autumn so far as bird-watching was concerned, though the usual scattering of Hoopoes and other small migrants was late arriving and rather thin on the ground. But in late April, a Purple Heron was closely followed by at least three Night Herons who made their temporary home in the Lower Moors on St Mary's. There had probably been a much larger arrival of Night Herons initially, as an islander who went to Porth Hellick early one morning reported having disturbed eighteen small herons from the sallows by the pool!

Then in early May I received a phone call from Francis Hicks on St Agnes to say that a Squacco Heron had appeared by their Big Pool. Next day it was on St Mary's and settled down there eventually for the next few days, mainly frequenting a small roadside pond at Newford where it found many small eels to feed on. Towards the end of its stay it took to sharing the grazing rights with sheep in a nearby field, but it was the flies swarming about the sheeps' droppings that were the main attraction.

Two Avocets, real off-beat rarities in Scilly, were seen on St

Agnes and Bryher but these were rather overshadowed by the various unusual herons, which increased in variety when I had brief but conclusive views of a Little Bittern among the reeds of Tresco Great Pool. But for me the greatest thrill was to enjoy the sight of a female Red-footed Falcon which I watched hunting tiger beetles on Tresco's Castle Down. There was also a considerable arrival of Quail that year, with birds calling on all the islands. It was certainly a spring to remember and I don't think there has been one to match it since.

Photography had to some extent taken over from painting as my medium of expression and although I still continued with the stylised Puffins and Oystercatchers that had become my stock in trade, it was purely to supplement my other earnings and no longer a pleasure to be churning out the same sort of pictures one after another. In order to satisfy the demand for this sort of thing I hit on the idea of producing black and white postcards, which I had printed from scraper board originals, and I found that these too sold very well at a few pence a time. Although there was not a lot of profit from this it saved me from repetitious work and so I concentrated more and more on building up a collection of colour transparencies to illustrate my slide-shows, which continued to pack in the customers on Saturday nights.

The Seabird Specials, too, were proving to be as popular as before and mercifully the boating war had subsided for a while. I guess the 'opposition' had realised by then that there was business for all and, despite the success of *Buccaneer*, the Association were still doing very nicely, so an uneasy truce ensued. Safari business was slowly on the increase though I found people very loath to commit themselves in advance, and was still very much at the mercy of the weather. However, the signs were encouraging for the future and I knew I would just have to keep plugging on if I was to make anything of my independence.

Now that I had been generally recognised as the most active bird-watcher in Scilly, it was only natural that I should have been invited to be on the editorial committee which had been set up by the Cornwall Bird-Watching and Preservation Society, to produce an annual report from Scilly. This had previously been included in the Society's main report as a separate section, but following the closure of the St Agnes Observatory, and consequent cessation of their biannual report, it was felt that a separate publication relating to the birds observed in Scilly was needed. The two co-editors were Miss Quick of St Agnes, to whom I referred earlier, and A.G. Parsons, a well-known ornithologist from Cornwall who had been visiting Scilly regularly for many years. Parsons was an eccentric in the grand tradition who not only wore old-fashioned clothes and used pre-First World War optical equipment but spoke in a high-pitched, pedantic fashion. Nevertheless he was an excellent observer and a meticulous collector of information on a wide range of subjects. I never got to know him too well although I met him on many occasions, but I had a great respect for his abilities. He helped me to obtain a good second-hand set of Witherby's *Handbook* and an improved telescope, both at extremely favourable prices, so at that time I felt reasonably assured that I was also accepted by the Cornish ornithological hierarchy. Although I did not have much of a hand in its production, the first *Isles of Scilly Bird Report* was published in 1970, though it of course related to birds seen in 1969 – it was gratifying to see so many of the records attributed to myself; it also gave me the opportunity of appearing in print as author of a short article entitled 'Scilly Honeyseekers' in which I demonstrated that it was not just as a recorder of rarities that I wished to be known, but also as a student of bird behaviour. Although I cannot claim that I had made any new discoveries it is interesting that now, fifteen years after this piece was first published, nothing new on the subject has come to light, at least if it has I am not aware of it. An abridged version of my article follows:

> It is common knowledge that the temperate climate of Scilly allows many plants from the Southern Hemisphere to flourish. Less well known is that in their natural habitats many of these are pollinated by birds. In coastal districts of Australia for example, Banksias, Eucalypts and others grow in dense stands which attract swarms of migrant birds which habitually feed on the nectar which gushes from the bottle-brushlike flowers.
>
> Although most of the trees from Australasia introduced to Scilly have only arrived within living memory, several of our common birds have learned to take advantage of this liberal honey-flow.
>
> The phenomenon is best observed in Tresco Abbey Gardens, where the rich variety of flora provides an almost constant supply of nectar throughout the year.
>
> The most spectacular example is undoubtedly the New Zealand Flame Tree (*Metrosideros*) which flowers profusely in mid-summer. These trees are sometimes smothered in scarlet blossom which attracts rowdy bands of Starlings from every direction. Sparrows, finches and even tits are also habitual imbibers.
>
> Earlier in the summer, one of the main attractions in the gardens is the tall flower-spikes of the Puja, a plant from South America. Each spike is covered with golden cup-like flowers, which after rain or a heavy dew are brimful with diluted nectar. Blackbirds particularly are attracted to this tipple, the birds' heads often becoming so dusted with golden pollen that they are mistaken for species of much greater rarity!

This was the first time that I had had anything published, other than bare factual descriptions of rarities, for a long while and although no literary masterpiece it was readable. I felt pleased that I had discovered another string to my bow which could be put to some use in the future – an ability to write the occasional article, although even then I probably realised it would not bring much in the way of financial reward.

My musical activities really took off that summer, and I found myself playing the trombone more evenings than not. Together with a varied assortment of friends with an even more varied assortment of musical abilities, we formed the nucleus of a band that could play in almost any of the local establishments according to requirements. At the weekly Folk Club for example, which in those days was held in the Mermaid, we were the 'Factory-packed Machine-bulked India Tea Jug Band' – the name derived from the tea-chest on which our bass-player strummed a single string accompaniment to our shaky vocals in the skiffle idiom which I had so much despised in my purist days! Another outfit comprising roughly the same people, though now playing proper instruments was 'Red Sails and the Sunset Syncopators'. I was Red Sails, and fronted the band on trumpet as well as trombone (though even with my versatility I was not able to play both at once). We played a mixture of trad jazz and outmoded dance music once a week at the Sunset Restaurant on St Mary's quay. And, finally, 'Preacher Deacon's Blues Band' in which I had a minor role to play, accompanying (inaudibly most of the time) a largely electric group consisting of too many guitarists with too little talent!

62. *'Red Sails' and the Sunset Syncopators*

The Adult Birdman Emerges

We did have one very talented guy in the band though, and I guess without him none of those musical aggregations would have ever performed more than once in public. His name was Dave Earl. Dave was an American guitarist who also played good banjo. Rooted in the traditions of American popular music, he seemed to be able to play almost any tune in any idiom and, what was more, knew the lyrics of every imaginable song from Stephen Foster to Chuck Berry. Unfortunately Dave was also a moody and very unpredictable character so inevitably the group's performance was as good or as bad as he himself was feeling on the night. On good nights we really had a marvellous time, but on other nights it could be very frustrating. The only other members of all those groups who had much to offer were Johnny Washboard (I forget his real name) who later went on to become a professional washboard-player for as long as it lasted, and Chris Garratt, an art student who was in the habit of coming to Scilly to work as a kitchen porter each year with other friends from college. Chris seemed to be able to play most instruments badly and could have been good on them all if he had applied himself, having a natural musical ability which often goes with other artistic skills, but it was only a bit of fun as far as he and his friends were concerned so we none of us took the music seriously and concentrated on enjoying ourselves. And that we indubitably did.

Inevitably I met a few girls in the course of my evening's music-making, and sometimes this led to a minor affair which helped to overcome the loneliness I suffered at being separated from Marianne and the children. Usually the girls involved were young enough to be my daughters and were no substitute for what I really needed, but I suppose it helped me to get my own back in a perverse sort of way; it certainly boosted my ego which had been severely dented by what I regarded as Marianne's disloyalty.

In addition to the popular Seabird Specials, we had started to do a once-weekly evening trip to Annet in *Buccaneer*. This we called the Shearwater Special, because besides the ever-popular Puffins, which could almost be guaranteed during the breeding season, there was also every chance of being able to show visitors the Manx Shearwaters which usually collected off-shore most evenings close to their main breeding colony. Although not quite pulling the same number of people as the day trips it was a useful addition to our repertoire. Of course it was not long before a rival trip commenced – they say imitation is the truest form of flattery, but I have to be honest and report that we were not even amused.

One boat-trip that we will never be allowed to forget happened that year when the tables were very neatly turned on us, and it was entirely our own fault. It was one of those days when the visibility was sometimes poor and occasionally lousy. Against our better judgment we were persuaded to take out the *Buccaneer* for a Seabird Special, even though we knew we would not be likely to see much. I guess we did not want to lose the chance of making a bit of extra money – always an important argument in favour of a boat-trip! As we were off the southern shore of Tresco the fog became thicker and in attempting to avoid a ledge that suddenly loomed up in front of us we struck a rock just beneath the surface which holed the *Buccaneer* sufficiently for it to be necessary to run her up on to the beach. The passengers got wet feet as they abandoned ship, but the only harm done was to the *Buccaneer* and her

63. *Shearwater Special*

masters – especially as we then had to enlist the help of an Association boat to get our passengers back to St Mary's. We have not made the same mistake ever again!

I still kept in touch with the family in St Ives as best I could that summer, and when Nicholas successfully passed his eleven plus exam and was accepted for the Humphrey Davy Grammar School in Penzance I played the part of concerned father (which I was) by going to an open day held for prospective new pupils and parents. There I was happy to detect a slight softening in Marianne's attitude towards me and even felt that it might be possible that if I could only get us a flat on St Mary's for the winter, she and the children would join me as soon as her season in St Ives came to an end.

That autumn was again relatively dull as far as rarities were concerned and, certainly from my point of view, did not match up to the spring which had been so exceptional. Perhaps it was because I had developed a number of interests apart from bird-watching that I managed to miss the greatest rarity of the year, a Scarlet Tanager which was seen by a handful of jubilant day-visitors from St Agnes, in some trees quite close to Porth Hellick Pool. I remember rushing to the scene only to find a scatter of disconsolate observers milling around the area in vain – it was never seen again. I had better luck with a Blackpoll Warbler (I had missed the previous one in 1968) on St Agnes a couple of weeks later, but the only bird

The Adult Birdman Emerges

worthy of note that I myself found that autumn was an Alpine Swift which for the whole of one afternoon hawked insects, in company with Swallows and martins, over a clearing on Tresco's Middle Down. It was not an auspicious autumn and I certainly did not live up to my reputation as a bird-finder, but I had things on my mind that were far more important because Marianne had made it known that if I could find suitable accommodation she and the children would join me on St Mary's.

Despite it having been a poor autumn for rarities, more and more bird-watchers were descending on Scilly in pursuit of them, and among the faces in the Mermaid, where the nightly gathering took place, was one that I seemed to recognise from television – I felt sure it was the little chap who sang daft songs on a programme called 'Twice a Fortnight', which in retrospect I think was much too often. Anyway, I accused him of this and he reluctantly admitted it was indeed he, but would I please keep it to myself because in Scilly he just wanted to be another bird-watcher, and not a TV freak! HE should have been so lucky.

Bill and I found out that we had quite a lot in common for besides our interest in birds we enjoyed jazz and rock music, had quite similar backgrounds, and enjoyed the same sort of jokes (well some of them anyway). This marked the beginning of a friendship which has continued through the years, and may carry on, subject to what sort of foreword he writes to this book! In case you are still wondering who the hell I am talking about, I can only assume that

64. Some familiar faces?

you have either gone through life permanently tuned to the radio third programme, or been on an intergalactic voyage for the last decade. Of course I am referring to W.E. Oddie, the noted ornithologist, council member of RSPB and occasional Goodie, among whose dubious talents are playing the saxophone, tending garden gnomes and compiling volumes of notes on the finer points of identification of *Phylloscopus* warblers. And that is not all!

There were of course many other faces which have since become well known on the Scilly scene, among them those of Peter Grant, Dave Holman, Paul Dukes, Harry Robinson and Ken Osborne to name only a few. In fact there were over sixty contributors credited in the *Scilly Bird Report* for that year despite the dearth of rarities.

7. Settling into a New Habitat

It was not difficult to find a decent flat for the winter and by the end of October we were all together at last in one of the basement flats looking out over St Mary's harbour. It was a great relief to be all under one roof again. The children very rapidly adapted to life on St Mary's: Nicholas started straight away at the new secondary school instead of the mainland grammar school he had gained entry to, while Martin rapidly made his mark at the junior school at Carn Thomas. Things were getting back to normal, but of course I now had to forfeit the bachelor life I had become quite accustomed to and settle down to becoming a responsible parent and a respectable member of the community – it was not easy. However, the temptations of the summer and the bird-watching season were both over and I was really very happy to resume the married state, and so I hoped was Marianne.

During the winter that followed I went back to work for Barry Mumford on the farm, and in my spare time rehearsed for a new role as the 'dame' in the local pantomime in which I once again got the chance to play the trumpet and sing 'I'm the Queen of the Jungle'. How I ever lived down that moment of madness I'll never know.

The big problem that now confronted us was finding a permanent home. Once the holiday season came round again we knew we would have to vacate our temporary accommodation, so somehow we had to get into alternative housing of some sort by Easter. I had already applied to the Isles of Scilly Council, without success, but felt that now I had the family with me I would have a better chance from that direction. But once again, though showing sympathy for our situation we were met by another refusal from the Council.

It was depressing in the extreme to think that after the winter together we might have to split up again for the summer. In desperation I agreed to take a flat approaching the full holiday rental in order that we could stay together – I didn't know how I was going to find the money, but somehow I believed if we could only get a place, that it would all work out for us in the end. And so it did, proving my optimism to be well-founded, for at the eleventh hour we were offered a chalet at a modest rental in exchange for caretaking a holiday cottage at Pednbrose, McFarland's Down, near the St Mary's Coastguard Station.

The chalet was little more than a converted garden shed cum greenhouse, but it became our home for the next few years, and if we had not been offered the arrangement by the owner, Miss Betty Astbury, I would probably be relating a very different story. Despite the very cramped conditions we salvaged what was left of our possessions from Barry Mumford's shed where they had been stored since leaving Tresco and made ourselves at home. With just two rooms, a kitchen-sink and toilet, we learned how to survive each other's company very well during most of the seventies. Luckily we had quite a decent garden in which to spread out in fine weather and I was able to raise a few vegetables while the boys, who had not been able to have much in the way of pets until then, kept a succession of small furry animals. Against a certain amount of pressure I resisted requests for us to own a dog, though we did occasionally look after other people's animals when they went away on holiday. I have nothing against dogs if they are properly looked after but in our circumstances it did not seem fair on any of us to add to the already overcrowded household. What's more it would not have been fair on the dog. In my opinion too many people keep dogs for the wrong reasons and when Martin asked – 'Why can't we have a wandering dog like everyone else?' this helped to enforce my decision against having one.

I was not too happy about the hamster, which quickly got lost, the gerbils which had a nasty habit of eating each other, or the guinea pigs either, but I suppose it was one way for the boys to learn some of life's little realities. When the ferrets came along we all had a lot to learn, but I'm getting ahead of myself.

So there we were in our own little shed in someone else's garden with nothing else to worry about except where the money was going to come from to keep us all in the discomfort we had become so used to. Of course my earnings alone were undependable so once more Marianne had to step into the breach and find a job. This was no problem – there are always seasonal jobs going for people prepared to work, and Marianne soon found a willing employer in the shape of Tim Clifford, formerly manager of Tregarthen's Hotel, but now owner of the Kavorna, a one-time coffee bar which he was in process of upgrading into a more salubrious type of establishment which has since become a bakery,

Settling into a New Habitat

grocer's and novelty shop. So with that problem easily solved, it was up to me to get on with my own affairs and try to provide the extras through the various means I had discovered while on my own during the previous two summers.

To try and provide customers for my safaris on a more regular basis I was now offering weekly 'Wildlife Holidays', which I advertised in the RSPB magazine *Birds*. The response was slow but encouraging, and my first weekly clients came along in the spring of 1971.

Not only did I promise them their full quota of excursions for the week, but I arranged their accommodation as well if required. So I was not only finding myself customers, but providing hotels and guest houses with extra business at the same time. At last it was beginning to make sense, and not only the boatmen but hotel and guest house proprietors began to realise that my presence on St Mary's might even be an asset – the astute ones among them even paid me a small commission. With this weekly nucleus of clients I could then add to the daily groups with casual customers when the need arose, which in those days was most of the time, for although I like to limit the groups to a maximum of twelve participants I was lucky to get them, and often went out with less than half that number. However, I was reasonably satisfied with my first season in business, and hopeful for the future.

Another innovation of that year was the establishment of Nature Trails on St Mary's. Without going into all the administrative background, suffice to say that Peter Mackenzie, now retired from flower farming, had been appointed warden by the Duchy of Cornwall and with advice from the Nature Conservancy set up these trails, through areas which had been hitherto virtually impenetrable, in the Lower Moors from Rose Hill to Old Town, and from Porth Hellick to the main road with a later extension to Holy Vale. Access to these marshy areas gave both local inhabitants and visitors the chance to get a closer look at birds, wildflowers and other natural phenomena, thus stimulating further interest and an awareness of what such habitats had to offer. In addition, two basic observation hides were constructed to give concealment for watchers at Porth Hellick Pool. All this was very laudable and I was quick to take advantage of it with my parties. Although I was invited to become a member of an advisory committee by the Nature Conservancy, which incidentally also included Lloyd Hicks and was chaired by my old pal the Commander, Mackenzie remained rather snooty about my activities and emphatically pointed out at one of the meetings that the hides were not intended for what he called 'large commercial parties' – I should have been so lucky!

Outstanding bird visitors that April included a couple of misplaced White Storks, which divided their time between St Mary's

Settling into a New Habitat

and Tresco, and a splendid Great Spotted Cuckoo which frequented the slopes between the airport and the sea, where it feasted on an emergence of Burnet Moth caterpillars. I was the autumn, though, which again came up with the goods as far as rarities were concerned.

A Lesser Golden Plover, still in summer plumage, appeared on Tresco in August but was rather overshadowed by an Avocet which was well watched at Porth Hellick for a few days around the same time. However, the real rarity, overlooked by the majority of people gawking at the Avocet, was a Wilson's Phalarope which though often in the same field of view was largely ignored! Then to everyone's astonishment two more White Storks appeared, standing on the tower of the St Mary's Church. This caused a temporary traffic hold-up, but after an hour or so, when everyone had seen them, the excitement died down. Still there the following morning, but circling away eastwards later in the day, they caused further astonishment when they stopped for another rest – on the roof of Penzance employment exchange! Because they had been ringed in Denmark, the movements of these Storks were well documented and one of them eventually got to Madeira later in the month.

Undoubtedly the rarity that caused maximum excitement that year was found on St Agnes where a regular gang of bird-watchers still continued to stay, despite the Observatory having closed some years previously. Luckily for me I had chosen to go there without the encumbrance of a group that day, hoping to see the Bonelli's Warbler which had caused a lot of controversy, almost culminating

65. *St Mary's first-ever traffic jam*

in fisticuffs earlier in the month, which was of course October, recognised among *aficionados* as the peak period for rarities. A rumour of an odd 'nightjar' was going around, but nobody took it too seriously until a cry came from someone, I think it was Peter Lansdown, that a NightHAWK had been seen and had dropped into an area of farmland which was out of bounds. Someone, who shall be nameless, cautiously trespassed the area until the bird had been located, whereupon a full-scale invasion took place. There seemed no way of holding it back and I soon joined the stampede, leaping over a low wall and hurtling after the muddy heels of a dozen or so flying twitchers! The Nighthawk was sitting almost invisibly on a little bit of ploughed land in a small field, resembling nothing more than a plain cowpat, so well was it camouflaged. Luckily I had the camera with me and took a few shots – just as I was attempting a closer picture the bird flew, but not without allowing me one quick shot of it which when processed showed clearly the main identification feature which I described at the time as 'broad white patches which extended across the primaries, and shone out with a luminance akin to headlamps' (what imagery!) That picture has been published several times in ornithological journals and I think it is probably the only photograph that ever earned me any money – perhaps enough for another roll of film!

The remainder of that year was spent mainly in consolidating my position as the recognised local bird-expert, building up the wildlife holiday business and trying to establish myself in the eyes of the

66. *A lucky shot of a Nighthawk*

local people as an asset rather than a liability. I realised that the early years of squatting with the Brooks family, trombone playing in the pub and like activities had not stood me in good stead with the establishment, although to the man in the street I might have appeared as acceptable. Having produced a wife and children gave me a little respectability, but to many my boat-trips, safaris and slide-shows did not add up to a proper job, and if measured by earning-power they were right! But I was if nothing else determined to prove them wrong in the end and continued to succeed with the visitors, who enjoyed not only my regular weekly programme but also occasional RSPB film-shows which were put on in the Town Hall, often drawing a full house. I introduced many holiday-makers to the work of the Society and raised modest amounts of money at the same time. Of course I was accused by detractors of using the name of the RSPB to promote my own ends, and to some extent it was true – the only status of any kind I could claim was as RSPB representative, but it seemed a very reasonable exchange to me, and there was no direct financial spin-off. Anyway I knew I had full backing from Trevor Gunton, the RSPB development officer, so I was not too worried on that score.

Trevor at that time came down as leader for a tour run by a rival organisation, Ornitholidays, who used to send groups to Scilly in both spring and autumn. That spring I remember both his group and mine enjoyed views of a splendid Gyr Falcon which desported itself about the islands for a few days in May. On the last occasion we saw it perched on the rocks of Stony Island as we passed by in one of the launches.

Trevor was also partly instrumental in helping me to set up a tour of mainland RSPB groups in the Midlands and London area during the winter of that year. Giving my standard slide-show 'Birds of Scilly' I was not only able to fill in weeks when I would otherwise have been idle, but also promote interest in my wildlife holidays. The scheme proved to be mutually beneficial to all parties concerned, and it established a pattern which I was to continue for several years subsequently during the seventies. It also of course gave me the opportunity of getting away from the islands for a while at someone else's expense, something I would have been quite unable to do without financial assistance. Although the lectures did not bring me in much if anything in the way of an income, at least I was occupied and doing something that was obviously appreciated by the audiences I met. It was almost like being 'on the road' again, as during my days in the music business, except that I did not have to worry about paying other people wages or whether I would have a complete band or not, and there were not the same temptations to go boozing or worse. At last I felt that David Hunt, Birdman, had really arrived!

During my lecture tour in the London area, I took up Bill Oddie on his invitation to visit him at his home in Hampstead, and even stayed there for a couple of nights. I met his wife Jean (like Marianne not an enthusiastic bird-watcher) and his two delightful daughters Katie and Bonnie. They all made me feel very much at home there, and I have been taking advantage of Bill's hospitality ever since. I hope he has managed over the years to get something in return.

I also found myself doing one or two TV jobs about that time. Tony Soper came over to record some material for a programme as part of one of his popular series, and my help was sought. Of course Puffins were one of the essential ingredients needed for a film about Scilly's birds, but as it was well into August by which time the Puffins are usually long gone, this was rather a tall order. We filmed various other species, but seemingly there was no substitute for the Puffins, so as a last resort we arranged a special evening trip to Annet, hoping there might be a few left, with a remote chance of filming a few Manx Shearwaters as well. In order to capitalise on the trip, we spread the word around that TV cameras would be on board, with the result that we filled the *Buccaneer* with visitors all hoping to see themselves on the box at a later date! There was only one problem – a westerly gale was imminent and the wind had already freshened considerably. This did not daunt the TV hopefuls, so off we set come what may. Of course we did not see any Puffins, but so what! I did my usual commentary as if they were there and the BBC archives supplied the Puffins in the final version! What's more the cameramen did get the hoped-for shots of shearwaters, so in the end it proved to have been a successful venture. Just to complete everyone's enjoyment of the evening, the heavens opened and we were all drenched with rain as well as spray by the time we got back to the quay. That's show business!

That was the first of several associations I have had with Tony Soper over the years – another was when he was doing a series about beachcombing. He contacted me well in advance with a list of natural and unnatural objects he wished to find; I then scoured the beaches for several days collecting what I could find, and when Tony turned up with the camera crew, my finds were all spread along the deserted beach of Samson for Tony to 'discover' in front of the camera. They included half a Sperm Whale, goose barnacles, a mermaid's purse and a message in a bottle! It's always fun working with Tony – he's so full of surprises.

Another TV programme I did was for producer Richard Brock about islands in Britain, from Scilly to Shetland – as usual I was called in to do the Scilly bit. This involved showing the viewers one of our unique Scilly Shrews, which I had to hold in my hand while nonchalantly explaining its uniqueness to the camera. I don't know how many people have ever actually seen a Scilly Shrew – I probably don't see more than a couple a year during my normal comings and goings (except dead ones). They are quite common, but being so small and inclined to hide away during the daytime, they are hard to find even if you want to. So our first problem was to find our shrew. We set a number of specially designed traps for catching rodents alive, but at first all we could catch were rats and mice. Eventually we managed to catch a few shrews as well, so filming could commence. I am not accustomed to holding small rodents and the first one I handled escaped at once – I was afraid of hurting it. We tried again and the next one I grasped more firmly – it bit me and scampered free to join its little friend among the boulders of the beach where we were filming. Richard Brock was not amused. 'You'd better hang on to the next one,' he said grimly – 'we'll soon have none left.' We eventually filmed the sequence successfully, but Richard never asked me to introduce a programme again. Next time you see Attenborough or Bellamy at work remember what they probably had to go through to get just a few seconds on film. And how often they have probably been bitten.

1971 also saw the founding of a Young Ornithologists' Club on St Mary's – drawing from pupils at both schools, we soon had the nucleus of a group, and of course Nicholas and Martin were among the early members. I had never attempted to push my children into becoming bird-watchers – if anything my obsessive behaviour while we lived on Tresco should have put them off for life! So it was pleasing to find they both supported the YOC from the start; together with some of their school friends we had a few outings among the islands and during the winter took part in the RSPB Beached Birds Surveys. Lewis Stephens, headmaster of the junior school, helped out with leadership duties, and since he knew all the children and their parents this of course made the job a lot easier – there was still the feeling among some islanders that bird-watching was a kind of madness, and I was the head loony, so having the headmaster's support was a big advantage.

It was with members of the YOC that we had the greatest thrill of 1972 – watching a Snowy Owl which appeared on St Martin's at the beginning of March. I received a phone call from one of the locals while I was on duty as an auxilliary coastguard look-out. He told me that what he described as a 'great white owl' had appeared there and was frequenting the area of Pernagie and White Island in the desolate north-eastern part of the island. It was several days before we could get over there but on the Sunday we scrounged a lift in a boat taking the minister over for chapel, and had just the duration of the service in which to find our bird. It was exactly where we were told it would be and the children and I had some marvellous views of it before it was time to hurry back to the boat. Like the previous Snowy Owl it turned out to be a long stayer, moving about the islands until late April.

Settling into a New Habitat

The coastguard job was a useful way of supplementing my meagre earnings during the winter. It involved a series of four six-hour shifts, with twelve hours between each shift, about twice every ten days. Since the shifts were often at night, this still allowed me to pick flowers most of the time when required, and the two part-time jobs gave me roughly the equivalent of one full-time pay packet each week. The biggest snag was that the hours involved often did not give me much time to sleep – for example, I could finish a night watch for the coastguard at 6.00 a.m. and have to start work on the farm at eight. My next watch then started at 6.00 p.m. Of course I could then get to bed at midnight but would have to be back on the farm next morning, work until twelve, and then back to the coastguard duty! There were days between when I had neither coastguard nor flower-picking to do, but I could not then catch up with the sleep I had lost. I have never been able to sleep well during the daytime, even when it was a routine, as it had been in Germany, and the cumulative effect of all this was that I used to fall asleep regularly while on night coastguard duty – luckily there was never a real emergency during any of my watches!

One of the other activities I also got involved in at this time was writing – I was asked to do an article by John Gooders for a magazine called *World of Birds*. Following the publication of this and regular pieces in the *Scillonian* magazine, published quarterly in those days, I embarked on a more ambitious project in collaboration with local photographer Frank Gibson – an illustrated guide to sea and shorebirds, followed by a series on wildflowers of Scilly. The latter were written mostly while I was on night coastguard duty – a good way of staying awake! The booklets also helped to establish my 'expert' status.

In May of the same year while conducting a safari to St Martin's, I had brief views of a very puzzling bird. About the size and shape of a thrush, it was a dull uniform grey-brown above and paler below, and without any striking features. It burst out from cover of a rock on a bare hillside, flew across a small bay and settled on a short grassy slope, where I watched it briefly as it stood with its wings drooped for a moment before dropping out of sight among some rocks. Although I attempted to find it again my efforts were unsuccessful and the search abandoned.

Next day was free, so together with Bill Oddie who was paying a spring visit for a change, we went in search of my mystery bird once again. Having failed to find it at the previous locality, we carried on along the coast until coming to the grassy area known as the Plains.

'Just the place for a Tawny Pipit this,' I remarked.

'Yes, and there it is!' replied Bill.

Sure enough, what could only be a Tawny Pipit was striding about on the short turf, wagging its tail for all it was worth, as if

Settling into a New Habitat

appearing by magic at the mention of its name! We forgot about my unsolved mystery for the time being as we enjoyed watching the Tawny Pipit, but I have often harked back in my mind to that puzzling bird. From what brief views I had, and judging by its behaviour and the locality it seems likely that it was a female Blue Rock Thrush. But because of the status of that species in Britain – so far not recorded as acceptable for admission to the British List except in Category D (probable escapes) I did not consider it worth submitting.

Although relations between myself and the boatmen had become reasonably relaxed after the initial campaign following my arrival on St Mary's five years earlier, it became clear to me in the early part of 1973 that hostilities were still not quite over.

I was quite taken by surprise to read in the letters column of the *Cornishman*, the nearest Scilly has to a local newspaper, the following piece, under the heading of 'Conservation No Protection':

> Sir,
> It was with a smile I read in your last issue about more protection for island birds, from the Isles of Scilly Nature Conservancy Committee. These gentlemen never seem to hit the nail on the head. Building nature trails and making hides do not protect birds, but help to drive them away.
>
> Before the nature trail in Higher Moors we had several ducks nesting at Porth Hellick Pond. Not now, and before David Hunt came along with his bird trips two or three times a week, with public address systems blaring, we had plenty of terns nesting on Merrick Island, Green Island and Stony Island, but not now. Stop the bird trips around these nesting places.
>
> Do away with the nature trails and the birds would be a lot happier. And to help the smaller seabirds, declare war on the Black-backed Gulls, because in twenty years or less there will be only Gulls, Shags and Cormorants left.
>
> G.P. Hicks (Boatman)

I was not only surprised, I was furious, for it seemed I had been singled out for attack as if I were some arch-priest of the disturbers, whereas the boatmen were represented as holier-than-thou nature lovers whose only concern was for the birds! Naturally I was moved to reply and this I did at great length, though the editor of the newspaper successfully emasculated my efforts to strike back by revealing what I believed to be the true facts of the matter by omitting the personality angles.

For your amusement only and not because I still want to get back at G. Hicks, now deceased but fondly remembered by many islanders and visitors alike as normally a quiet chap and the most unlikely person to indulge in mudslinging in the press (I still wonder who put him up to it), I now include the full version of my reply, minus the boring details of breeding statistics, population changes and other arguments I put forward to refute his case.

Sir,
Your correspondent Mr Hicks may smile as much as he likes at the efforts of our committee to protect island birds, but I personally find it no laughing matter that, in this enlightened day and age, anyone can still hold such old-fashioned ideas or be so totally blind to the facts. Actually, one does not need to be very clever to read between the lines of his letter and detect certain undertones which clearly indicate some ulterior motive behind the hypocrisy of the views claimed.

Since Mr Hicks has singled me out for special mention, I feel it is up to me to answer him and make a few points of my own. [Then followed an explanation of the present situation with regard to breeding success of ducks at Porth Hellick, and a statement of numbers recently observed there.] Perhaps Mr Hicks has not seen them because he normally does not go near the place without his dog, and with a gun under his arm. Could it be that he is suffering from a dose of sour grapes because we are trying to protect what he would prefer to find on his dinner plate?

Mr Hicks suggests that 'blaring public address' systems (and incidentally he has had one of these installed on his own boat for some time) have driven the terns off various islands. [Then followed the details about natural fluctuations in tern colonies.] Since the islands in question are designated bird-sanctuaries, it is a great pity they do not choose to nest on them every year!

Finally, Mr Hicks suggests waging war on Black-backed Gulls to help the smaller seabirds. Could it be perhaps that he would like an excuse for some legalised bloodshed with rifle and shotgun? Since this is unlikely ever to be approved, nor is it an efficient means of control, he will have to look elsewhere for his 'Sport'. Or perhaps he has found a new one – 'conservationist-baiting'.

David Hunt (naturalist)

A.G. Parsons also took up the challenge in his characteristic long-winded fashion, but presented very reasonable arguments for the existing conservation methods and there the matter rested.

No doubt if the paper had printed my reply in full there would have been further repercussions, so in retrospect I should have been grateful for what at the time I regarded as editorial bias in favour of the boatmen's view. Anyway there was no further obvious bad blood between me and 'G' Hicks to follow, so I guess it all ended to the satisfaction of both sides. Actually I too was among

'G's' admirers mainly because, unlike his brother Lloyd, he kept his mouth shut most of the time.

I even attempted to bury the hatchet with Lloyd Hicks who now accepted my existence as a *bona fide* naturalist if not an equal. We agreed to combine our talents and make the evening trip to Annet a joint venture rather than a hustle for business. With Lloyd as skipper I joined the *Swordfish* as crew and commentator and we shared the profits – you'll have to guess who took the largest share! In fairness to Lloyd, we had some good trips together and he was always friendly and easy to get on with; but in the end, because our two personalities did not really complement one another, it was mutually decided that one season was enough and the following year I resumed with *Buccaneer*.

That autumn was the dullest I can remember, probably because settled weather without the westerlies (which usually bring the transatlantic vagrants) prevailed throughout September, and even in October the only American bird was another Upland Sandpiper which by a strange coincidence (or was it?) settled into the same field on Peninnis as had the one in 1968.

In November I embarked on quite an extensive mainland lecture-tour, mainly to RSPB groups, but also to a few schools, among them Ravenswood where I was the guest of the headmaster Richard Mackie who, incidentally, had been head boy on my very first arrival there! I also paid my first visit to Gresham's and renewed acquaintance with Dick and Pat Bagnall-Oakeley. Many things had changed in the twenty years since I had left Gresham's, but I'm glad to say I no longer felt the same way as I had done on leaving in 1952. The new headmaster, Logie Bruce-Lockhart, was a bit of a birdman himself, and made me feel very welcome. I also had a fleeting visit to Cley, reviving old times with Dick and even finding time for a chat with Richard Richardson. Sadly, I was never to meet either again, as both died prematurely before I next visited Gresham's. It was not until that final meeting that I came to appreciate fully what a profound influence the two had been during my developing years.

Early 1973 was as uneventful as had been the previous autumn and apart from the odd Hoopoe and Golden Oriole there was little to excite my holiday clients apart from the resident species and breeding seabirds. Just the same, business was improving, thanks to the combined results of advertising and my winter slide-shows, so I had every reason to feel satisfied on that count. Sometimes I was even in the position of having to turn people down who wanted to join safaris at short notice.

Fine weather during the main part of the summer greatly benefited the sea and shore birds' breeding season that year, and it was good to see the efforts of our local conservation committee's policy for protecting the tern colonies on Tresco taking effect, with many

Settling into a New Habitat

young birds in evidence in July.

Meanwhile, the children were growing up at a dramatic rate. Nicholas, in his third year at the Isles of Scilly School, was showing a keen interest in music; he had even taken up the trombone (which was the only available instrument in the house) to play in the school orchestra; and Martin was proving to have quite a talent for golf, spending much of his time practising on the nearby course, where he received a lot of encouragement from the steward Harry Wright and his wife Mabel. Marianne continued to work at the Kavorna, where she had become virtually indispensable. Everything was going fine in fact, apart from our quest for a proper home.

This need became even more pressing as the boys grew older and bigger. Nicholas's musical tastes were widening to include folk (he now possessed a guitar) and rock music. The chalet often shuddered to the repetitive riffs of Led Zeppelin or strained under attack from the ear-splitting vocals of Sparks, two trendy groups of the period. How Marianne's nerves put up with the aural assault I don't know, but Martin and I survived by being out most of the time, me on business, and Martin on the golf course.

An initially unwelcome addition to the menagerie at that time was Tramp, a baby ferret which Martin had found abandoned and brought home in hopes of being allowed to keep it. He first as usual asked his mother. 'See what Dad has to say when he comes home,' she replied, guessing that I would be against it from the start.

Thinking it would be lucky to survive the night, being barely weaned, and to avoid a scene I told Martin he could keep it if he looked after it himself. I fully expected that it would be dead next morning. But Martin somehow coaxed it to survive, fed it milk from his fingers, took it to bed with him at night and before we knew it Tramp had become one of the family! As she got stronger and bolder (we only found out later she was a female) we became more and more attached to the animal despite her smell which soon pervaded the chalet to such an extent that we no longer noticed it. Martin showed a side of his character through his persistence with Tramp that we did not know he possessed, and were glad of this because apart from his golf and other sports he seemed uninterested in doing much to better himself, especially at school.

Having served five full years as editor of the *Isles of Scilly Bird Report*, Hilda Quick finally bowed out in 1973, leaving the task to me and a new recruit from the previous year, Harry (HPKR) Robinson, a Scilly regular who had recently moved from Hampshire to live in Cornwall. Parsons had somehow fallen out with the Society, many of the younger members regarding him as a living anachronism, and although none of my doing had taken umbrage, regarding my appointment as editor as a personal snub. I was sorry about this, having had a lot of regard for him as an ornithologist; following his death in 1982 it would be inappropriate not to state that his help and encouragement during my early years in Scilly was much appreciated – there will never be another AGP.

So now I was not only Scilly's leading field ornithologist, I was editor of the *Bird Report* and official recorder (the two jobs went together) as well. Without doubt I was now indisputably the 'Scilly Birdman', and until someone comes along prepared to relieve me of those duties, and believe me they are not a labour of love but of necessity, I seem destined to remain so.

In our 1973 *Bird Report*, Harry and I endeavoured to bring the style of the publicaton in line with those of other regional and county reports by including all the species occurring in the islands, rather than just selected migrants and rarities. We also started to include detailed descriptions of the rarities that merited such treatment, with drawings if available. Luckily we were funded by the society, or this would not have been possible. I like to think that these innovations were a success, especially since I have continued the formula ever since.

For the drawings in our report we have been lucky to be able to count on many of the best illustrators in the field of bird-artistry. Expertise of the highest quality has come from Peter Grant, a stalwart from the days of St Agnes Observatory who, despite his rise to the dizzy heights as Chairman of the BBRC, is still a regular devotee of the 'Scilly Season' and always finds time to encourage the younger generation of up and coming birders. Ken Osborne provided the excellent drawing of Puffins, which graces the front cover of every issue, and is an eye-catching advertisement for it. Richard Millington, who earns his living by portraying birds in water-colour, is never too busy to send a few drawings and there have been many more artists who have willingly contributed over the years. I cannot mention them all.

But there is one more worthy of special mention and that is Bryan Bland whose work, you cannot fail to notice, adorns many of the pages of this book. Bryan's is a unique talent, and I am very grateful to him for what he has done in the past for the *Scilly Bird Report*, but especially for his efforts to portray many of the events that have been described less masterfully in my text. His work speaks for itself.

Recently, Mike Rogers took over from Harry Robinson as co-editor of the report, and has done a great job by relieving me of the main burden – the systematic account of birds seen through the year. As Honorary Secretary of the BBRC he is the ideal person to do it, and long may he continue in both roles! Mike has talked for a long while about coming to settle permanently in Scilly. If he does maybe one day I can hand it all over to him!

8. Leader of the Flock

Looking back on the mid-seventies it is hard to recall the sequence of events too clearly, as I had really established a steady life-style as a naturalist, with the emphasis on birds, though I was also able to hold my own in allied subjects. As a result the only things that help to distinguish one year from another are the birds and to a lesser extent the people involved in finding them.

1974 I best remember because of a momentous boat-trip, accompanying a shark-fishing expedition with skipper Johnny Poynter in the *White Hope*. Bill Oddie, Andy Lowe, Robin Hemming and myself in between being bored (Andy), vomiting over the side (Bill), beating the sharkers at their own game (Robin) and falling between two cameras (myself), we saw Leach's Petrels, a Sabine's Gull and a Long-tailed Skua, to say nothing of other assorted seabirds. There were other rarities that autumn, including Scilly's first (and only) Sharp-tailed Sandpiper, two Solitary Sandpipers, a Paddyfield Warbler and a Yellow-breasted Bunting. And that was just the extreme rarities!

67. *The momentous boat-trip*

The people who spring to mind from that year are Robert Allen, who came to Scilly ostensibly to do a seabird survey for the Nature Conservancy, but eventually stayed on and off for the next five years, Dave Flumm, already a Scilly perennial who contrived to spend most of the summer and autumn here and John Yrizzary, the American who first dared put a name to the Sharp-tailed Sandpiper! Dave Flumm also stuck his neck out by claiming the identity of the Paddyfield Warbler, and was eventually proved right after suffering much scorn from those who labelled it as a mere Sedge Warbler. What a year that was – but even more amazing years were to follow.

Robert Allen was an ex-university man with (I think) a degree in English but a passion for birds. He had approached the Nature Conservancy about the possibility of doing some seasonal work and had been offered the job of surveying Scilly's seabirds single-handed! I was first introduced to Robert by Peter Mackenzie, who seemed to

68. Robert Allen

Leader of the Flock

think well of him, despite his unruly tangle of golden curls and generally anti-establishment image. Robert was wary of me at first (I later discovered that Peter had warned him to be careful of me, though in what way I was never quite clear) but we soon became good friends and enjoyed many happy birding experiences over the next few years. His final report to the Nature Conservancy showed that he had put in an enormous effort in assessing the breeding status of all the local seabirds, together with a lot of research into the food of nesting gulls. Unfortunately Robert was not encouraged to continue working for them for various reasons, partly self-induced, so most of his initial research has been wasted through not having been followed up.

The first real rarity Robert discovered was a Solitary Sandpiper, which he disturbed from a tiny brackish pool on the remote little island Rosevear. It flew off in the direction of St Agnes and disappeared. Hearing of his find as soon as he returned to St Mary's, I decided to search every likely spot on the island. To my great surprise and delight, it was teetering on the muddy edge of the first small pool I looked at, and I was to see it again next day on the Great Pool, Tresco. If there is ever good luck in birding I must have been leading a charmed life that year, for I missed nothing, and even got some reasonable photographs of several rarities.

At home, 1974 is best remembered by the somewhat belated arrival of the Beatles! Nicholas had not been interested in pop music at the time of their first emergence upon the British public, but Martin decided to make up for this more than a decade later. I suspect there was some devious subconscious reason for a clash of musical taste between our two sons, but it was never more apparent than during that summer. Nick had become so proficient on the guitar that he was already an integral part of a local folk-group calling themselves, not inappropriately, Crow Sound (in fact the name of a channel between the islands). Apart from his tastes in pop, which had long transcended the Beatles, Nicholas listened to a lot of the emerging electric folk-music as played by groups like Fairport Convention and Steeleye Span.

So every day when school was over, there was a tussle about whose turn it was to use the record-player, and for the rest of their waking hours we were being serenaded by either 'She Loves You, Yeah, Yeah, Yeah', 'All around my Hat' or similar sleep-precluding ditties. But it was the Beatles who usually came out on top, due to Martin's qualities of persistence and Nick's reluctance to indulge in physical violence unless provoked beyond endurance. Somehow we all lived through it, and so did our neighbours, who with four kids of their own to contend with contributed in no small way to the din.

Luckily Martin's passion for golf and Nicholas's various outside interests meant that there were times when neither of them were at home, but I rarely got to appreciate them because I was out most of the time myself. As usual it was poor Marianne who had to bear the brunt of it!

A certain measure of relief from this situation came when Nicholas, who had done very well in his 'O' levels, went away for his further education to a Sixth Form College at Brockenhurst in Hampshire. And then Martin, whose golf had improved so well that he had been entered for a junior tournament, spent a while with his old friends Harry and Mabel in Yorkshire. The relief was only of short duration.

Martin, on his return from holiday brought with him what he described as a nice surprise – a large and boisterous male ferret! Hobo, as he soon became known, was quite the opposite of Tramp, not only in gender but in temperament. However, they had to live together in one cage just as we did. Nicholas invited some of his new school friends to come and camp in the garden as well, so summer 1975 is one that lingers in the memory more than the average one.

That autumn is equally memorable, but on quite different grounds, for 1975 was also the 'Year of the Sapsucker'! But before the arrival of that milestone in Scilly's ornithological history, there were other visitors from America, the first one causing me not a little embarrassment. This was a medium-sized wader which a visitor who was staying on Tresco reported to me as a Greater Yellowlegs. I went to investigate, and found the bird quite approachable – I was even able to get a series of photographs. But with my vast experience of the species (none) I scorned the identification, stating that the bird was definitely a Lesser Yellowlegs (of which I had

69. *The Greater Lesser Yellowlegs*

previously seen two). Nobody disagreed with me, and several other observers saw it during the next few days, until it was found dead, partially eaten by some predator. I obtained what was left of the yellowlegs, which included the head, legs and a wing, and just for the record measured them – to my astonishment they fell well within the range of variation in the Greater, but were far too large for Lesser Yellowlegs. So my Tresco informant, C. McCullough, had been right all along in his original identification. I was suitably chastened and learned a maxim that I still occasionally forget – never pronounce a bird's identity unless you have first checked its credentials. There is another maxim I would also recommend after that *faux pas* – never trust an 'expert' they are only human after all.

Then came an avalanche of records of American birds, starting with possibly as many as eight Buff-breasted Sandpipers, another Solitary Sandpiper, a Lesser Golden Plover and a dowitcher. And on the mainland there were similar occurrences, culminating in the arrival of a Black-billed Cuckoo in, of all places, Yorkshire! Besides this we heard that Fair Isle had had Britain's first ever Tennessee Warblers – it was incredible.

When on 26 September the leader of another tour-group visiting Tresco told me they had seen a Wryneck in an area of sallows close to the Great Pool, I decided to take my own group and see if we could find it too, as there was not much about. It was not long before a likely-looking bird showed itself.

'There's the Wryneck,' I exclaimed – 'No, wait a minute.' I gave the bird a second, closer scrutiny. It was a woodpecker, but clearly not one I was familiar with, especially on Scilly, where any woodpecker is a remarkable event (I've still only ever seen one, a Great Spotted, also on Tresco). This bird was a real puzzle, and I asked the members of my group if they had any ideas. As it turned out, most of them being virtually beginners, no one had any suggestions – the poor fools expected me to have all the answers! Usually I have, but on this occasion I was stumped, even then I had not even considered it being an American vagrant, believing that no woodpecker could have the faintest chance of crossing the Atlantic.

So I decided that for the moment the priority was to get as detailed a description of the bird as possible. This was not too easy because it was mainly in silhouette against the light, and I did not want us to scare it by going too close. Eventually during the course of the afternoon, together with a few stray observers I rounded up (the other group had gone back to St Mary's, contentedly thinking they had seen a Wryneck!), we assembled enough plumage detail on paper to satisfy me we could later identify it from the literature.

That evening I pored over my books and after eliminating all the European possibilities I disbelievingly referred to my American field-guides. I was soon in no doubt that our bird was a Yellow-bellied Sapsucker, but could I really claim it as such? After my experience with the Greater Yellowlegs I did not want to end up with egg on my face again. But conviction that on this occasion I could not be wrong prevailed, and that night the news spread rapidly across Britain through the twitchers' grapevine. Next day there was a mad scramble for helicopter seats at Penzance, and those who could not get one took the morning steamer. Wisely I decided that I would stay away from Tresco and leave the business of confirmation to the newly arrived experts. No one disagreed, and during the next ten days or so hundreds of birders descended on Tresco to see it. It was a worrying time for the Tresco estate – they had never experienced anything like it, and the gamekeeper was sent, gun in hand to control the hordes of invading twitchers, some of whom attempted, and succeeded, in sleeping rough on the island without being noticed. Bird-watchers had already made themselves unpopular on St Agnes by thoughtless behaviour in the past, and it became evident that somehow I was going to have to start a movement among the bird-watchers themselves, to control the bad behaviour of the small minority. This was another problem.

To make matters worse, or from the birder's point of view better, another 'crippling' rarity turned up a day later on St Mary's, but the news did not get out until Saturday night, when I was shown drawings of a bird which was clearly a Black and White Warbler. Next morning at daybreak a crowd had already gathered in the fields close to the pines on St Mary's Garrison, where the warbler had first been seen, and after an anxious half-hour of searching it was finally spotted, high up among the pine branches. It also stayed around for a few days and was seen by many people. Apparently the chap who first found it had thought it was an Azure Tit, not having had an American field-guide to refer to – it was rapidly becoming required equipment for any birder visiting Scilly!

This became even more apparent the very same day when Roy Alderton and Graham Hearl discovered an immature male Scarlet Tanager in the pines at Gimble Porth, Tresco. It was all too much – three amazing birds, all from America, and within as many days! It did not end there for in October Dave Holman, a young birder just beginning to make a name for himself, discovered a Bobolink in a weedy field on St Mary's Peninnis Head. And then a Surf Scoter appeared in the channel near Tean, where I ran a special boat-trip in *Buccaneer* in order to see it – the only time I ever made any money out of a bunch of twitchers! Finally Scilly's third Blackpoll Warbler arrived.

But it still was not over, for when the weather eventually changed, the invasion came from another quarter. The easterly winds brought at least fifty Yellow-browed and three Pallas's Warblers, several Richard's Pipits, Rustic and Little Buntings and two immature Rose-coloured Starlings (the first since my Bryher

70. Products of an easterly wind

bird ten years ago). The most remarkable of all these eastern vagrants was a Bimaculated Lark, a bird not normally found further west than Turkey!

My personal favourite, though, was the Little Bustard and the story of that bird is worth relating. First seen on Agnes by Tim and Carol Inskipp (at that time she was Miss Robinson), it was pursued on to the Gugh from where it flew back towards St Agnes and disappeared. It was sought high and low on all the islands for the next few days, including Annet and Samson, but no trace of it could be found. It was presumed either dead or departed, October came to an end, and only a handful of visiting birders remained. Among these were Roger and Liz Charlwood, who were searching for the elusive Rustic Bunting at Trenoweth Farm, St Mary's, when they flushed the Little Bustard from an overgrown bulb-field. It flew off and disappeared in the direction of Telegraph. Roger (and I am eternally grateful to him for this) had the presence of mind to phone me – remarkably I was at home and since the bird had disappeared very close to our chalet, we searched each field with caution. Sure enough it was in a former potato-field on the corner of the lane, but it flew off again before we could get a proper view. This happened again later, and I even got a photograph of it in flight, but never saw it on the ground. Imagine my disgust when Charlie Mudge, a neighbouring farmer, told me he had seen it in his fields quite regularly, and Betty Blackwell, who runs the airport buffet, said it was often on the runway first thing each morning.

A final anecdote from that momentous autumn also involved Roger Charlwood. Together we were walking across the gorsy heathland on St Martin's, and I commented 'Not likely to see much up here except a Dartford Warbler', whereupon one fluttered across the path and gave us some conclusive views. I had never seen one before in Scilly and have not seen one since – such is the power of suggestion!

1975 marked the beginning of what I can only call the annual 'Twitchers Invasion'. The quest for rarities which first started in the days of St Agnes Observatory had now reached an unprecedented high, but it was a trend that was only to increase during the next few years, and from which for a while I did my best to escape.

Having thoroughly established myself as a wildlife tour leader and Scilly's leading birdman, the struggle to survive seemed a thing of the past, even though we could not find a home of our own. I had become quite conditioned to living in the chalet, and with Nick away most of the time, the pressures were not as marked as they had been for a while. Nevertheless, I did feel sorry that Marianne had to suffer the cramped accommodation situation and consequently began to feel increasingly guilty. Prices of property had escalated so rapidly that to buy a house was out of the question, but I continued to nag at the Isles of Scilly Council, though without success – there was always someone in more desperate need than us, and anyway we were making a go of it so why worry seemed to be the attitude. So we had to carry on as we had done for the past five years with little hope of ever getting a place we could really call our own.

After the excesses of the previous autumn, it seemed very unlikely that autumn 1976 would produce an even greater influx of bird-watchers, or as they were increasingly being called 'birders', following the American trend for listing, which by then had really caught on in Britain. Rather than face another season of madness, I decided that I would no longer conduct parties during the month of October, leaving it free for me to do my own thing or even go away if I could afford it! I was becoming increasingly envious of the tales I heard of trips to India, America and elsewhere and determined that, if I could get enough money together, October would be the time to go on holiday, and leave Scilly to the twitching masses!

So after a moderately successful season, Marianne and I between us saved up enough for a week's package holiday in Majorca! No very big deal you may think, but it was the first holiday we had enjoyed together (apart from occasional visits to my family, once to Germany for Christmas) since our trip to Yugoslavia in 1959! We both enjoyed it – me for a chance to go bird-watching on my own in new surroundings – Marianne for the chance to laze on the beach in the sun, which paradoxically she rarely has the chance to do in Scilly.

While on holiday in Majorca, we met Eddie Watkinson, RSPB rep for the island, and his wife Pat. Eddie had written a very popular guide for bird-watchers on holiday there, and told me that the publishers of his book were looking for authors for similar guides. It seemed like a good idea to write one for bird-watchers visiting Scilly, so I brought back not only fond memories of our holiday from Majorca, but plans to write a book.

71. Some of the Grey-cheeked Thrushes

I had not been back in Scilly more than minutes before I began to meet birders with tales of the rarities I had been missing during my self-imposed exile in Majorca. I was soon caught up in the madness once again, for if 1975 had been the year of new varieties, 1976 was fast becoming the year for quantities of American vagrants! Two Black Ducks had been discovered on Tresco which I wasted no time in getting over to see, then no less than five Grey-cheeked Thrushes, two Rose-breasted Grosbeaks and perhaps as many as eight Blackpoll Warblers, not to mention the American Robin on St Agnes that appeared during the latter part of October. And to think I had been to Majorca to get away from it all!

An incident which still gives me some amusement dates from that October, and it involves my old friend Ron Symons. Ron, in the quiet way so typical of him, casually mentioned one day that he had found a dead American Nighthawk on the beach at Porth Hellick.

'It's in my garden shed if you want to see it,' he said.

I repeated the story to some of the birders, who seemed to be singularly unimpressed.

'Probably that Nightjar that's been roosting up in Holy Vale – it's disappeared recently,' said one of them.

Knowing Ron's bird knowledge to be better than that of the average twitcher, I said nothing, but decided to go and see the dead bird for myself. Ron was having his tea when I arrived.

'Go and help yourself, it's in the shed – I don't want it,' said Ron, who was not prepared to interrupt his meal for a dead bird. Sure enough, it was a genuine Nighthawk, and I took the corpse down to Holy Vale to show it to some of the scornful birders. Of course a dead bird does not count as a 'tick' on the twitcher's list and since few of them had been around at the time of the St Agnes Nighthawk several years previously, most of them 'needed' this one. I often wonder whether it was the Nightjar which had been roosting in Holy Vale. Another dead Nighthawk was found soon afterwards near the airport, but the birders had to wait until 1981 for a live one.

And so the pattern continued for the next couple of years. I would go away for a week or two in October to 'escape' for a bit, and as soon as I got back was thrown once more into the hurly-burly of the 'Scilly Season' as it had now become known to the ever-growing army of birders. I slowly became more accustomed to it, and as I did so realised that like it or not my presence in the islands was needed, especially when I heard tales of birders misbehaving, fights between them and island youths, and other unsavoury stories. So in the end I resolved to stay in the islands during the annual invasion, to try to act as not only a buffer between birders and residents, but as spokesman, vigilante and general ornithological ombudsman.

With the coming of the 1980s, I found myself having to act in quite a different role to that which I had embarked on back in 1969, when I first had visions of becoming the Scilly Birdman. Then I had been quite content to devote all my energy towards self-preservation and establishing myself as an individual, in order to reinstate my family in the local community. Now I was having to consider much wider issues, for with my final acceptance, aided by the recently published *Guide to Bird-watching in the Isles of Scilly*, the local people were looking to me to advise them on how to encourage or control bird-watchers!

The first step in the right direction had been made back in 1977 when we published, after consultation with various local organisations and individuals, a code of conduct for visiting bird-watchers. This had helped, not only by drawing attention to the various aspects of birder behaviour most likely to find disfavour with islanders, but also by making local people aware that there was a movement among the birders themselves to put things right.

As each year more and more birders were arriving at the very end of the season, boatmen, publicans and proprietors of all kinds of accommodation were beginning to see that birds meant business, and I was regularly asked advice on how to encourage bird-watchers. In some ways that has had a bad effect on my own business, but more about that later.

The most important happening for us towards the end of the seventies had been the offer of a new job for Marianne, whose abilities were at last recognised for what they were worth. The offer came from Roy Mitchell, the local building contractor who had recently built a complex of holiday cottages at Porthcressa in Hugh Town St Mary's. As well as a job with a proper salary managing the holiday cottages, she would get one of the brand new cottages for us all to live in. It was as they say 'an offer you can't refuse' and I would have been the last one to stand in her way, so finally we were to have a decent roof over our heads. Ironically Nicholas was no longer with us, having left school and gone to live in London in hopes of finding fame and fortune in the world of music! Still, with or without him it was great to have a spare room and a bath of our own. And we still had the ferrets!

So, in the autumn of 1978, we moved in to the brand new premises of 16, Silver Street – we had become town-dwellers again rather than country folk! Since the town is smaller than the average mainland village, this was no great hardship. We were once again within a stone's throw of the beach, and I would only have a two minute walk to the quay and boats next season. At that time we could only appreciate the advantages of our new situation, and if there were any drawbacks were not prepared to think about them.

In any case I was soon off again on a lecture tour, but on this occasion not to the conurbations of industrial Britain but to Sweden, where I had arranged several engagements as well as some free-time diversions through my publisher friend in Stockholm, Jeremy Sanders. This proved a success on all counts, and I realised then that there were future possibilities of lecture tours, perhaps even in America if I played my cards right.

On my return from Sweden I was saddened to hear of the death of Hilda Quick, whom I had got to know as a good friend and ally since taking over from her as editor of the *Bird Report*. I attended her funeral in the little church on St Agnes in all sincerity, although I generally avoid such commitments if I possibly can. At the same time I was able to catch up with the 'rarity of the year', Britain's first-ever Semi-palmated Plover, which was still lingering on the beach at Periglis, only a few yards from the church! I know Miss Quick would have approved.

72. *Semi-palmated Plover (left) attending Hilda Quick's funeral*

Without the problem of finding a regular winter income any more I was able to afford the luxury of giving up the flower-picking that year, but found that the less energetic life-style resulted in a rapid increase in weight and feeling of unfitness. To counteract this, in common with people all over the world at that time, I took up jogging or, as I prefer to call it, running. Without wishing to break any records or make too many demands of my reluctant body, I did not have a very punishing routine, just enough to keep the muscles and tendons working so that when my safari season started once more, I would be in a position to lead from the front as usual.

My normal route took me out and around Peninnis Head, usually returning to Porthcressa by the opposite path to that which I'd run out along. This offered a slight change of scenery, because I'm sad to say that the beauty of Scilly's land and seascapes does cease to have the same effect on one, after twenty or so years, that it had on first acquaintance. It also gave me a chance to look out for any odd birds that might appear, especially during spring or autumn migration periods. Naturally my binoculars always accompanied me on these occasions in case I saw a bird which needed more than a glance, or to provide me with a good excuse for a breather!

One morning in late May 1979, I was on my customary jog around Peninnis, musing upon the possible birds that a clear dawn with light southerly breezes might have brought in. Knowing that most of the migrants had already arrived, I was not left with many likely species, though of the unlikely ones Rock Thrush came to mind as being a remote outsider, and at the same time highly desirable, for not only would it be a new bird for me but for Scilly as well.

So when a bird resembling a Rock Thrush flew up from the ground to perch on one of the massive boulders that are littered all over the headland, I at first thought I must be hallucinating. But a second glance confirmed my first impression, and when I finally

73. *Rock Thrush – reality from an illusion*

got my binoculars de-misted (have you ever tried to use them after you've been running for half a mile?) I became quite sure that my illusion had in fact become reality, and that in front of me was an immature male Rock Thrush. It was not unduly shy, and I was able to watch it as it moved around, building up a full description. Then I remembered that Robert Allen, who at that time was still trying to settle on the islands, and like me ten years earlier had found a temporary home at Number 6, North Parade, would also like to see it, and would be the ideal person to corroborate the record (though by then I had no fears of having my records rejected by the BBRC). Robert was still in bed but he shot out when he had convinced himself I was not fooling, and soon he too was able to enjoy the spectacle of that exciting rarity. Being a Saturday, there were also a few lucky birders over for the day from Cornwall, but after that it was not seen again. Oddly enough, when Bryan was on Peninnis in 1984 sketching this scene, another Rock Thrush turned up on the very same rocks.

I continued my morning run until that autumn, but nothing to match the Rock Thrush appeared. Then one day I had not gone very far when I heard the sound of another runner coming up behind me, but he was no fellow-jogger – rather a young twitcher, who asked in a strong Liverpudlian accent 'Am I running for a good reason?'

'Yes,' I told him, 'because you're a bloody fool.' He had thought I was running for a rare bird!

Running for rarities is not something I usually indulge in, though I do remember once running down from Telegraph Hill in order to see an Alpine Accentor, also on Peninnis Head, but that was a long way and I didn't know how long it was going to stay. (It stayed for about a month!) However it is a trend that I find is becoming ever more prevalent among the younger and more enthusiastic birders. I applaud their enthusiasm and I admire their energy, but I sometimes feel that they carry it too far. I felt like kicking in the teeth of the fools who ran up behind me when I was trying to photograph the Black and White Warbler, and scared it immediately into the background.

Recently I was walking out again towards Peninnis to investigate reports of a Lesser Short-toed Lark, when I was overtaken by a gentleman, not young either, who asked me why I was proceeding so purposefully. I told him about the hypothetical rarity.

'Oh is that rarer than a Short-toed Lark?' he asked.

The answer to that you would not know unless you are either a very knowledgeable ornithologist or a fanatical twitcher.* But this chap was obviously neither, so why was he heading for Peninnis in such a hurry? He didn't know, and I still don't, except that he had obviously been caught up in what I can only describe as the

*Answer: Yes, the only previous records were in Ireland in 1956 and 1958.

'October Madness', which affects quite ordinary people who come to Scilly at that time of the year.

If I may give another example of the same sort of thing: two experienced looking lads, both with standard 'Barbour' waterproofs and heavy boots, are standing outside the Porthcressa Restaurant, where I have an information board detailing the recently reported rarities and where they have been seen. The list of sightings is pretty amazing and includes several super-rarities, but also a few other birds of lesser rarity, among them a 'Ring-tail', which as any experienced birder knows is short for immature Hen or Montagu's Harriers, which are not easily separable in their 'ring-tailed' plumages from adult females.

The two young men are looking puzzled as I walk past them, so I stop and ask them if they have a problem.

'Yes,' says one, 'this ring-tail – we've looked in both the European and American field-guides, but can't find any mention of it.'

Now this is forgivable from beginners, but the point I am making is that beginners should not be chasing around the Isles of Scilly in October. There seems to be no solution to the problem, because no one can deny youngsters the chance of coming to the islands, and God knows we need the business; if only they would somehow spread their attentions out into other months of the year!

To return to my narrative, a year later Roy Mitchell acquired the premises of Porthcressa Beach Café, which he rebuilt eventually as Porthcressa Restaurant, the idea being to provide fast-food service in the daytime and à la carte by night, for holiday-makers and residents. The basement of the premises became available for functions, including film-shows and discos, while the top floor was converted into a modern flat for the management, which as you have no doubt guessed by now was to be my wife! Inevitably I became involved too.

In the autumn of 1981 the still partly uncompleted cellar of Porthcressa Restaurant was opened to visiting birders in the evenings, in order to provide them with an information and entertainment centre where they could eat, drink and talk birds to their hearts' content. The venture was immediately successful, and the regular slide-shows, a nightly log, presided over by BBRC secretary Mike Rogers, and an annual 'Birdwatchers' Ball' (come dressed as your favourite bird) have become an integral part of the 'Scilly Season'. And I of course have become the genial host (well, some of the time).

Not that I have discontinued my other activities, far from it, but with the coming of acceptance by the local community as well as the visitors, and the financial stability provided by Marianne's new job, I have been able to relax rather more than in those early years, and even be more selective about the work I do. For instance I have done two successful lecture tours in America where, although I did not earn a fortune, at least I paid my way! And I have become deeply involved as an overseas tour leader for Cygnus Wildlife Holidays, which has taken me five times to India, to West Africa and the Pyrenees. And will hopefully take me to many further exotic destinations in the future. Who knows, one day I may have enough material to write a sequel to this book, entitled 'Travels of a Scilly Birdman'.

But it is with my Scilly career that I am dealing in this book, and it is with some of my adventures and encounters with people during the course of my ordinary routine that I will be concluding in the chapter which follows.

9. Adventures in the Bird Trade

With all those tales of rarity hunting and twitchers, the average reader might be excused for thinking that my life is one long round of excitement and rich with incidents, but of course for most of the year I am engaged in guiding very ordinary people, often complete beginners, and only very occasionally experts, but all of them needing my assistance in seeing the wonderful range of flora and fauna that the Isles of Scilly have to offer. During the course of these activities we do occasionally stumble upon the ultimate rarity, as with the famous sapsucker, so I must always be on my toes, prepared for whatever good luck occasionally throws my way. But for the great majority of the time my work involves leading small groups (I try to limit the size of the group to a sensible maximum) in pursuit of what for me is the ordinary, though fortunately for visitors often something totally new and wonderful. Every year, for example, I show hundreds of people their first-ever Manx Shearwaters and Puffins, and I regularly find people who are as enchanted by a familiar Stonechat or common wildflower as others might be in some hyper-rarity. I am lucky that I have found a niche in life that allows me to indulge in my own hobby, while helping others to enjoy theirs at the same time – I hope I will never feel otherwise. There are times of course, as in any job, when I would be happy to do something else for a change – but when I think what that might be, I must admit I am hard pressed for an answer.

It would be strange indeed though if I did not have a few tales to tell about my holiday clients, so here are some of the nicer ones – I'll leave the others to your imagination! For obvious reasons I'll not be naming any names, but if you recognise yourself in what follows please don't be offended – it could not happen to a nicer person.

My first story took place on Tresco in the autumn of 1976, when a Bobolink had been reported the previous day close to my old

74. *A familiar Stonechat*

home at the Blockhouse. We spent most of the morning, together with roving bands of twitchers, searching the dunes and coastal scrub without success – the Bobolink seemed to have disappeared. When lunch time came, the group settled down with their sandwiches, using the Blockhouse Fort as a convenient picnic place – it also gave a commanding view of the area, should anything appear while we ate our lunch. I noticed in one of the derelict gardens below us that a little group of birders had assembled – they were all gazing into the same bush, presumably at a rare bird. Telling the group to wait and watch, I went to investigate – the bird turned out to be a Wryneck. Knowing that for several of my party this would be a new bird, I waved to them to come down and join me. As they descended the hill a bird, which turned out to be a Tree Pipit, flew across and settled on the TV aerial of one of the cottages. Since I knew that this would also be new for some of them, I shouted 'Look at the aerial!' They did, and after a brief look, because it soon flew off, joined me to find that the Wryneck had also disappeared. 'Never mind, the oriole was nice,' said one of the my customers. 'Oriole – what oriole?' I asked. It seems my injunction to look at the aerial had been misinterpreted as 'Look at the oriole'. The fact that the bird was a Tree Pipit had not sunk home! After all, anyone can make a mistake.

Adventures in the Bird Trade

I have also been guilty of misinforming my clients from time to time, and the best example of that was when one very wet and unpleasant day in October 1974 I was conducting my party, come what may, on St Agnes. I had heard that a Spotted Crake was frequenting the Big Pool, as it is laughingly known (only qualifying to be called big because the other pool is a puddle). When we got to the pool it was still tipping with rain, but that did not deter the crake which was feeding in full view at the edge of the mud. 'That's it, there's the Spotted Crake,' I said, barely giving it a glance, and left the group to look at it while I scanned the nearby beach for something more interesting. I did not find anything, and in the end we gave up for the day wet and bedraggled, the only consolation being that we had seen a Spotted Crake. A couple of days later, and after my group had finished their holiday, the rumour spread that the crake had now been reidentified as a Sora, and the first in Britain for many years (there have been two more in Scilly since). I wonder how many of my group at that time still have fond memories of a wet day on St Agnes when they saw their first Spotted Crake? Or if they were told the truth, would they know the difference, or even care? All questions I'm unable to answer, but without such people I would not be in business, for big listers are not among my normal customers. Luckily they usually prefer to look after themselves, though there was one person who hired my services, with just adding to his list as his main objective for visiting Scilly.

By some coincidence, it was shortly after the discovery of the above-mentioned Sora. My client arrived on the Saturday, a day I normally keep free for my own personal affairs. 'What's about?' said Mr H (I'll call him that for convenience). I told him about the Sora on St Agnes. 'Come on then, what are we waiting for?' he said. 'I'm afraid we've missed the boat now,' I replied. 'Well hire one then,' ordered Mr H. 'I'll pay whatever it costs for a special launch.'

Mr H obviously knew what he wanted and how to get it, for within a few minutes we were heading rapidly for St Agnes in a hired speedboat. Needless to say we did not see the Sora which had disappeared, since an attempt to catch it had failed the previous evening. But that did not bother Mr H because right where the Sora had been I discovered a Jack Snipe, which being new for his list was a perfect substitute – he did not seem to attach any 'value' to the birds he saw, each one being merely an addition to the list, upon which he started looking for another! On the nearby beach some Sanderlings were running up and down at the edge of the tide. 'What are they?' asked Mr H. I told him. 'Oh yes, I last saw those in Seychelles,' he replied. Clearly he was no ordinary customer.

During the week that followed I showed Mr H many new birds. He carried a clip-board with him, on which he wrote down each

163

75. *Tresco Myrtle Warbler*

bird we saw in sequence, no matter whether they were new for the list or not, and he wrote them as if in an inventory: Redshank – common 2, Wren 1, Linnet 4, Snipe – jack 1, etc. I have never met anyone like it since, and will be glad not to – it is not the kind of bird-watching that appeals to me. I see nothing wrong in keeping a list of birds, or even aiming for a target number of species per day or year or whatever, so long as the main objective is a love of the birds and the fun of bird-watching. But Mr H's preoccupation was with the list, which could just as well have been of cigarette packets or car registration numbers.

The clip-board did come in useful before the week was out though, because I dictated a description of a strange warbler we found on Tresco to Mr H and he wrote it down word for word as I spoke them. It turned out to be a Myrtle Warbler, and another for Mr H's list!

My next story relates to two of my favourite clients from the early days. The ladies concerned were regular safari sufferers for several years and although they come less regularly these days, I am and will always be pleased to see them because they know how to enjoy their birds, whether rarities or not, and nobody knows better than I do how impossible it is to produce rarities to order!

Anyway, my friends had gone to Tresco one afternoon on their own in pursuit of a mystery falcon that one of my other clients had reported and even photographed earlier in the week. They failed to see the falcon but came back pleased with themselves at having seen a fine male Snow Bunting in spring plumage. I believed them, it was not so unusual, but one of the ladies later retracted the

Adventures in the Bird Trade

76. *Snow Bunting or No Bunting?*

sighting in some embarrassment, explaining that she realised that the 'Snow Bunting' had in fact been one of the free-flying budgerigars that in those days infested Tresco. I could not understand how they could have made such a mistake, being quite experienced, but once more believed them and forgot all about the matter.

That winter I was staying with the man who had photographed the mystery falcon and asked to see the picture (it was a Lanner, and probably an escaped falconry bird). 'I've got a picture of that Budgie that posed as a Snow Bunting as well,' said my friend. You've guessed of course – it was a fine male Snow Bunting! And that is what I am up against!

Luckily with observers of that quality, I do get people coming back for more instruction which is just as well because in the end I would eventually run out of clients. In fact this has already happened in some respects, though some clients I never want to see again!

I should have known from the spidery handwriting that Miss X (I'll call her that to avoid libel) would be nothing but trouble, but I failed to read between the lines of the series of letters she contrived to send, just to arrange a holiday booking. I never attempted to help people to find their own way to Scilly, believing that it was just a matter of common sense, but somehow in the case of Miss X, I became involved in advising her on the relative merits of train versus coach, and boat versus helicopter. Actually all she really wanted to know was which would be cheaper, and anybody could have told her that. So she travelled by coach to Penzance a day earlier than she needed to, had to spend the money she had saved in the coach on overnight accommodation, then came over by boat. In those days I used to try and meet people off the steamer, though it was seldom a successful venture. I used to stand on the quay with a placard saying 'Wildlife Holidays' – it attracted plenty of remarks, but my clients, being observant types, invariably walked past me with a glassy stare, quite oblivious of me or my placard. Of course Miss X was bound to be one of those and sure enough the quay slowly emptied of people and she had not materialised, so I decided to go to her guest house and see if she had turned up there without my assistance.

The guest house door was open, so I went in, took a quick look in the lounge, where all I saw was what appeared to be a heap of old clothes, then knocked on the door of the kitchen to see if the landlady was at home. 'Miss X has arrived, she's sitting in the lounge,' were her first words. Funny I thought, and went back for another look – the bundle of old clothes had a shrunken head and a pair of spindly legs, I observed. Miss X had indeed arrived!

I did not see Miss X use her legs until that evening, when she tottered in, bent over a stick, to attend my Saturday slide-show, the introductory part of the week's programme. 'I'm rather deaf,' she said, 'where would you recommend I sit?' I did my best to get her seated near me, but I had doubts about how she would get on later in the week. The Sunday Seabird Special was no problem once I had arranged to get her on the boat, but when we arrived at Tresco on the Monday morning I knew I was really up against it!

The poor old soul was so frail that she could make little progress against the strong northerly wind that hit us as we walked up the quay at Carn Near, Tresco's southern arrival point. Somehow, with much patience from the rest of the group, and adopting a snail's pace, we managed to get as far as the Abbey Gardens, less than half a mile from Carn Near.

'Wouldn't you like to stay here and have a look around the gardens while I take the others to view the pools?' I suggested.

'I'll not be left alone,' replied Miss X. I was nonplussed, never having had to experience this kind of thing before. Then one of the other ladies in the party, who had been with me the previous week kindly offered to stay with her and my problems were over – for a while at least. We enjoyed the rest of the day on Tresco, and somehow Miss X got to the quay for the return journey.

The rest of the week presented further problems, but by offering alternative routes and substituting boat-trips or bus tours for the more strenuous walking, I thought I had at least managed to fulfil part of my commitment to Miss X and refunded her some of her advance payment because she had to miss the final excursion in order to catch her coach back to wherever she came from. I have seldom been more relieved to part with a client, feeling that I had done my best for her in the circumstances. I felt sorry for her really, because she had had little idea of what to expect in Scilly and my holiday programme, although not physically demanding, was obviously more than she could cope with (I later found out that she was over eighty?). And that was the last I saw of Miss X.

But she had the last word, for a week or so later I received a letter from her in which she claimed that I had not given her value for money. How dare I suggest she was not fit to participate fully in my excursion programme; because of me she had had to curtail her holiday, etc., etc. She went on to list all the other tours she had been on all over the world, including archaeological digs, botanical expeditions and so on. I did not bother to answer it, because I knew it would only invite further fruitless correspondence. I am glad such customers have been rare in my experience – mostly I hope my clients go away satisfied.

Sometimes I have upset people I never even met; that was the case with a certain French gentleman we'll call Monsieur G. I should first of all explain the rather complicated situation which sparked off the whole unfortunate affair, which I hope will read like the plot for a TV comedy by the time I've finished.

It all started one day on St Agnes in the spring of 1977. We had heard that a couple of Hoopoes were frequenting the area of Wingletang Down though had not so far seen them at lunch time, when we went as usual to the Turk's Head pub for a pint of beer and one of their delicious pasties (one of my weekly weaknesses). During our lunch break some friends of mine, a local couple with a mild interest in birds, came in and the subject of the Hoopoes came up. 'I'd love to see a Hoopoe. You know I've never been lucky enough to see one in all the years I've lived here,' said the husband. I promised to let them know if we were lucky during the afternoon and, sure enough, one of the Hoopoes duly appeared, flying up and disappearing from view in a grassy clearing among the gorse. Just at that moment, as luck would have it, I saw my friends approaching from the opposite direction. Unless I could warn them, they were obviously going to disturb it again as they came along the path. To complicate matters, most of my party had not yet seen the bird so I was also concerned that, if it was scared off, they might not see it either. There seemed to be only one solution – I shouted to my friends and tried to indicate by waving that the Hoopoe was very close to them. Seeing me, but not getting the point, my friends waved back and kept on walking. I shouted again, but too late, for the Hoopoe rose from the gorse and disappeared over the far side of the slope. Most of my group had missed it again! We hurried after the bird, all taking different paths through the gorse, but were not able to find it again, so gave up and continued our walk. Apart from feeling rather frustrated that I had been unable to show the

Adventures in the Bird Trade

Hoopoe to either the group or my friends, I gave the matter no further thought.

Several weeks later I received a letter from the Director of the RSPB, with an enclosure which read as follows:

Dear Sir,
During a recent holiday in the Scilly Islands I witnessed very strange practices in the field of ornithology. One could call it 'beating the bushes', 'scaring rare wanderers' or 'paramilitary birdwatching'. I call it genuine vandalism.

On St Agnes at the end of April there were two Hoopoes. One day as I was observing them from a hidden place I heard voices and other noises. I looked up and saw about fifteen people all wearing binoculars walking towards me in line. They recalled a sight I dislike very much from seeing it here in the South of France; a group of hunters and beaters determined to get their prey at any price. Of course the Hoopoes flew away. By then a whole concert of shouts took place and the line collapsed. Everybody started to rush and run after the birds, trying I suppose to get closer to a valuable tick. I left them to their silly games.

Of course I gathered local information about these methods and their organiser, a man called David Hunt who makes a living out of the birdwatching trade. Daily M. Hunt and his customers visit a different place. You know as well as I do how difficult it is to watch birds when you walk among a brightly coloured group, especially in bare landscapes. But when people have paid to be shown a Black Duck or a Yellow-bellied Sapsucker, you have to make the bird fly and worry it until everyone is satisfied. When I was told that M. Hunt is the RSPB man for Scilly I could not believe it until I read in your last issue of *Birds* an advertisement which confirmed this. I felt I should write to you and share my indignation. I still believe I was unlucky to cross the path of M. Hunt, because there must be very few of his kind among your staff.

That was not all, but I have included the meaty part of Monsieur G's letter to show how easy it is to give the wrong impression even when conducting a perfectly respectable tour group about the islands. Actually I am particularly concerned myself with the way bird-watching is sometimes brought into disrepute by the behaviour of bird-watchers and, since the Director of the RSPB was perfectly aware of this and knew there must have been some misunderstanding, I did not have to go into a lengthy explanation. However, I did sympathise with Monsieur G, and wrote to him at some length, apologising for having upset his day and trying to explain my own views on the matter. I never received a reply, and as far as I know he has never returned to Scilly. I do feel his reaction was rather extreme but one has to accept that, as with any other form of recreation, one meets a great variety of people in bird-watching and it is not possible to please everyone all the time. Someone once told me that she had overheard another lady saying she would not like to join David Hunt's safari party because 'He drinks!' And I thought that a mere pint at lunch time showed great moderation!

Sometimes I have been the one blamed for a misdemeanour on the part of my group; since I am the tour organiser this is fair enough I suppose. There was a time on Bryher when we had walked along a trail with a gate at each end which was required to be kept shut, as was clearly painted in large red letters on each gate. As I assume the people who come with me on safari can read, and in any case should know the basic country code, I do not like to continually remind the last person to shut the gate. Of course on this occasion the gate was not closed and a pony got loose. The irate farmer came chasing after us on his tractor and delivered the deserved abuse. In all innocence (I could not believe anyone in my group could have been so careless) I denied the accusation. Imagine how I felt when one of the party told me later that she had left the gate open because she did not want to get her shoes muddy! I saw the farmer later and apologised, explaining that I felt it embarrassing continually to have to remind people to shut gates. The incident was then reported in the Bryher column of the bi-annual *Scillonian* magazine thus:

Mr David Hunt and a swarm of bush beaters went through Mr Langdon's land and left the field gates open, not once but twice. On being tackled by Mr Langdon about the ethics on farm and country code, Mr Hunt's reply was that he found it embarrassing to have to ask his party to shut gates. I leave it to your imagination as to what reply Mr Hunt was given.

Yes, it's a small world – especially in the Isles of Scilly, but after twenty years or so one gets used to it. By setting myself up as the local expert I have invited gunfire from the local inhabitants from time to time so must be prepared to take it as it comes. One day perhaps I'll get shot down for good, but not without putting up a fair fight.

Which brings me to the subject of the future, and what it holds for me and my adopted home in the islands. My past record for predicting the future has not been remarkable, but it has always been based on optimism. That alone will not carry us through the eighties, in which we will have to mix optimism with a considerable amount of action. I hope to be fully involved in that action.

10. Present and Future Status

Twenty years have now elapsed since I made that first exploratory visit to Tresco, and fifteen years since I took the plunge by squatting on St Mary's, with the hope of one day becoming a fully-fledged birdman. Just when I finally reached that elevated status is not important, but it would be fair to say that I have now arrived. I have been described as a 'sort of ornithological Godfather', an 'adviser to twitchers and a guide to dudes' – I prefer to call myself simply 'the Man on the Spot'. And that is what I hope to remain for the rest of my working life. Not that I have any plans for retirement – in theory I retired when I left Tresco and started working for myself. But unfortunately I am a hard taskmaster and I don't provide an index-linked pension scheme for my employee, so it looks as if I'll be working for my boss until I drop!

Being the Scilly Birdman in the 1980s is not what it was a few years ago, but I suppose this is just another symptom of the world we are living in. From a business point of view, two major factors contribute to an overall decline. Firstly, the high cost of getting to the islands has meant a general dropping off in trade all round, and especially in the money people are prepared to spend on peripheral things like safaris – food and drink come first! Secondly, and this is linked with cost again, people who can afford an expensive holiday are weighing up the relative values of, say, a visit to the United States, or the Far East and finding they offer better value for money if not taken too often – and worth saving for. So in between these exotic holidays, they take shorter and cheaper breaks in Britain where it is now possible to attend weekend courses, wildlife holidays or whatever (again I am not mentioning any names) in almost every country or region – and certainly much less expensively than in Scilly. So now as one of the originators of wildlife holidays in Britain I am having to take a back seat. Of course Scilly has lost none of its attraction; in fact the facilities available now are

Present and Future Status

vastly improved compared to when I started, but the problem is not what we have to offer – it is purely and simply the high cost and difficulties involved in getting here.

Not that these deterrents affect your dedicated birder for whom money spent on travel is money well spent – provided of course there will be some good birding to follow. But this again narrows the field, for as every twitcher will tell you – October is the month. This is a most unfortunate situation for as more and more birders, many of them virtually beginners, are wanting to visit the islands they have heard so much about, it is always October that comes to mind. This results in far too many birders arriving in the peak season for rarities, and very few in the main migration periods of spring and early autumn, which can start in late August and carries on through September, a month which has produced many of Scilly's outstanding rarities if not the bulk of them.

What can be done about this situation? I wish I had the answer, but slowly, as more and more newcomers pile in during October I hear a steady murmur from the old hands, who are becoming disenchanted. Perhaps this will result in more birders coming in September to beat the rush, or in November to catch the latecomers – and the first weeks of that month have been consistently good during the last few years.

Spring is my favourite season and, over the years, as you have read in previous chapters I have met many exciting birds during that period. Unfortunately it is much less predictable than autumn, and a birder coming in April or May risks a dull time if the weather is unsuitable, although of course the usual sea and shore birds are always to be found, and the wildflowers are then at their peak. But I fear that the quest for rarities which has become so important in the minds of the majority today (and I must admit that I bear some responsibility for promoting it) will not die as rapidly as the twist or the hula-hoop!

So you may say, and I am sure many of you will, I have successfully fouled my own nest, and this is to some extent true. But I am the eternal optimist and feel sure that during the next few years the situation will improve, not only for me but more importantly for the islands themselves, and I sincerely hope to have a part to play in that. What that part will be is still very unclear in my mind, and depends on several factors over which I have little or no control, like the general economic situation and how seriously our elected government take their responsibilities towards our environment. But for what they are worth, here are a few predictions. Most of them are frivolous. The problem I will set you is to determine which ones.

Firstly, by the end of the eighties, Scilly will have declared UDI and will be functioning as an autonomous free state, with no taxes but communications subsidised by the EEC. There will be a strict

Present and Future Status

limit to the number of motorised vehicles allowed on Scilly roads, with a total ban on non-essential transport. Dog-owning will not be permitted without written permission from the dog!

In this ideal new island set up, there will be an indpendent Environmental Trust, in which local people, including David Hunt, will have full control over access arrangements to wildlife habitats, many of which will be designated sanctuaries. Full consideration will also be given to other interests, among them fishing and shooting, for which a new code of practice will have been worked out to suit all parties concerned.

Visitors to the islands will be encouraged by low-cost air and sea passages from Britain, and climate-control will have been established to give sufficient sunshine to attract holiday-makers all the year round; but there will also be nightly rain quotas to satisfy farmers, who will then be able to produce tropical fruits and blooms instead of boring old daffodils.

Bird-watchers (who will no longer be known as birders or twitchers, but as 'blisters') will have identity cards and ration books, allowing them a quota of five 'ticks' per visit, after which they will be required to leave the islands. First-time blisters will have to spend an indoctrination period on one of the less popular islands for at least a month before being allowed on St Mary's. After this period they will be issued their 'Qualified Blister' certificate which will entitle them to a card and ration book on their next visit. Only one visit per year will be allowed during October, though a second visit may be made at any other time.

Is this all a nightmare or a lunatic's pipe-dream? Remember it was written in 1984!

77. *That's all folks!*

Postscript*

'You're far too young to have written your autobiography.' So spoke a friend early in 1985. A few weeks later David Hunt died. The circumstances were almost unbelievable. He was leading a party of bird-watchers through the Corbett National Park in northern India. He spotted a rare owl and, warning the rest of the party to wait and stick together, he set off in pursuit of the bird. He followed it out of sight over a ridge. There, he was attacked by a Tiger and killed.

The ironies are numerous. Like all birders visiting India, David longed to get a 'decent view' of a Tiger. I, myself, have visited the country several times and have never even seen one. Friends have returned proudly with a fuzzy slide of a stripey tail disappearing into the jungle half a mile away. The odds against actually meeting one face to face – let alone being attacked – are millions to one. As a conscientious and highly responsible tour leader David would have been the first to blame himself for wandering off on his own. As a passionate conservationist he would have been the last to 'blame' the tiger. The reports in the Sunday papers would, I think, have amused him: 'Close friend of Prince Charles killed in India'. Well, David *may* have met the Prince for an hour or two in Scilly sometime or other ... I don't know; but – with all respect to HRH – I doubt if it would have been the highlight of David's week, let alone his life. The posthumous elevation to 'a close friend' would have tickled him. The fact is David didn't need false fame or fictitious friends – he was known, respected and loved by literally many hundreds of people.

Not much that David did in life could be described as 'ordinary', and so it was to the end. Considering the imminent publication of this book I have a feeling his own wry comment would have been: 'Anything for publicity!' I trust no one will find such a conjecture offensive ... having read his life story, you must now know that David certainly wouldn't!

When he died David was 51. He *was* far far too young. He was unique, talented and quite irreplacable. I, and many others, miss him very much indeed. Scilly will never be the same again.

Bill Oddie
March 1985

*Nothing in the text of the book nor Bill Oddie's foreword has been altered subsequent to David Hunt's death. We feel certain that is how he would have wanted it.